High-Energy Photons and Electrons:
Clinical Applications in Cancer Management

Wiley Series in Diagnostic and Therapeutic Radiology

Luther W. Brady, M.D., Editor

Professor and Chairman, Department of Therapeutic Radiology and Nuclear Medicine, Hahnemann Medical College and Hospital, Philadelphia, Pennsylvania

TUMORS OF THE NERVOUS SYSTEM

Edited by H. Gunter Seydel, M.D., M.S.

CANCER OF THE LUNG

By H. Gunter Seydel, M.D., M.S.
Arnold Chait, M.D.
John T. Gmelich, M.D.

CLINICAL APPLICATIONS OF ELECTRON BEAM THERAPY

Edited by Norah duV. Tapley, M.D.

NUCLEAR OPHTHALMOLOGY

Edited by Millard Croll, M.D.
Luther W. Brady, M.D.
Paul Carmichael, M.D.
Robert J. Wallner, D.O.

HIGH-ENERGY PHOTONS AND ELECTRONS:
Clinical Applications in Cancer Management

Edited by Simon Kramer, M.D.
Nagalingam Suntharalingam, Ph.D.
George F. Zinninger, M.D.

TRENDS IN CHILDHOOD CANCER

Edited by Milton H. Donaldson, M.D.
H. Gunter Seydel, M.D., M.S.

High-Energy Photons and Electrons: Clinical Applications in Cancer Management

Edited by

Simon Kramer, M.D.,

*Professor and Chairman,
Department of Radiation Therapy
and Nuclear Medicine,
Jefferson Medical College,
Thomas Jefferson University,
Philadelphia, Pennsylvania*

Nagalingam Suntharalingam, Ph.D.,

*Professor,
Division of Medical Physics,
Department of Radiology,
Jefferson Medical College,
Thomas Jefferson University,
Philadelphia, Pennsylvania*

George F. Zinninger, M.D.,

*Associate Professor,
Department of Radiation Therapy
and Nuclear Medicine,
Jefferson Medical College,
Thomas Jefferson University,
Philadelphia, Pennsylvania*

*Proceedings of an International Symposium
"The Clinical Usefulness of High-Energy Photons and Electrons (6-45 MeV)
in Cancer Management," Thomas Jefferson University, Philadelphia, Pa.,
May 22-24, 1975.*

A WILEY MEDICAL PUBLICATION

JOHN WILEY & SONS

New York / London / Sydney / Toronto

Copyright © 1976 by John Wiley & Sons, Inc.

All rights reserved. Published simultaneously in Canada.

No part of this book may be reproduced by any means, nor transmitted, nor translated into a machine language without the written permission of the publisher.

Library of Congress Cataloging in Publication Data:
High-energy photons and electrons.

   (Wiley series in diagnostic and therapeutic radiology)
   "Proceedings of a symposium held at Thomas Jefferson University in May, 1975."
   Includes index.
   1. Cancer—Radiotherapy—Congresses.
2. Radiotherapy, High energy—Congresses.
I. Kramer, Simon, 1919-   II. Suntharalingam, N.   III. Zinninger, George F. [DNLM:
1. Radiotherapy, High energy—Congresses.
2. Neoplasms—Radiotherapy—Congresses. QZ269 H638 1975]
RC271.R3H53   616.9'94'06424   76-10616
ISBN 0-471-50685-0

Printed in the United States of America

10 9 8 7 6 5 4 3 2 1

Authors

ADAMS, G.E., D. Sc., Ph.D., Gray Laboratory, Mount Vernon Hospital, Northwood, Middlesex, England

ALMOND, PETER R., Ph.D., Department of Physics, The University of Texas System Cancer Center, M.D. Anderson Hospital and Tumor Institute, Houston, Texas

BLOEDORN, FERNANDO G., M.D., Professor and Chairman, Department of Therapeutic Radiology, Tufts-New England Medical Center, Boston, Massachusetts

BRADY, LUTHER W., M.D., Professor and Chairman, Department of Radiation Therapy and Nuclear Medicine, Hahnemann Medical College and Hospital, Philadelphia, Pennsylvania

DUTREIX, ANDRÉE, Ph.D., Professor of Radiophysics, Institût Gustave Roussy, Villejuif, France

DUTREIX, JEAN, M.D., Professor of Radiotherapy, Institût Gustave Roussy, Villejuif, France

FEHRENTZ, DIETER, Ph.D., Universitäts Strahlenklinik, Czerny Krankenhaus, Heidelberg, Germany

FLETCHER, GILBERT H., M.D., Professor and Head, Department of Radiotherapy, The University of Texas System Cancer Center, M.D. Anderson Hospital and Tumor Institute, Houston, Texas

FOWLER, JACK F., D.Sc., Ph.D., F. Inst. P., Gray Laboratory, Mount Vernon Hospital, Northwood, Middesex, England

HALL, ERIC J., Ph.D., Professor Radiology, Radiological Research Laboratory, College of Physicians and Surgeons, Columbia University, New York, New York

HO, JOHN H. C., M.D., Consultant in Charge, Medical and Health Department, Institute of Radiology and Oncology, Queen Elizabeth Hospital, Kowloon, Hong Kong

JOHNS, HAROLD ELFORD, Ph.D., Professor and Head, Department of Medical Biophysics, Ontario Cancer Institute, Toronto, Ontario, Canada

KORBA, ALVIN, M.D., Fellow, Division of Radiation Oncology, Mallinckrodt Institute of Radiology, Washington University School of Medicine, St. Louis, Missouri

LAI, K.C., D.S.R., Medical and Health Department, Institute of Radiology and Oncology, Queen Elizabeth Hospital, Kowloon, Hong Kong

LAM, C.M., M.Sc., Medical and Health Department, Institute of Radiology and Oncology, Queen Elizabeth Hospital, Kowloon, Hong Kong

LAUGHLIN, JOHN S., Ph.D., Professor and Head, Department of Medical Physics, Memorial Sloan-Kettering Cancer Center, New York, New York

MANEGOLD, K.H., Ph.D., Max Planck Institute of Biophysics, University of Frankfurt Frankfurt, Germany

PEREZ, CARLOS A., M.D., Professor of Radiology, Division of Radiation Oncology, Mallinckrodt Institute of Radiology, Washington University School of Medicine, St. Louis, Missouri

POHLIT, W., Ph.D., Max Planck Institute of Biophysics, University of Frankfurt, Frankfurt, Germany

POWERS, WILLIAM E., M.D., Professor and Director, Division of Radiation Oncology, Mallinckrodt Institute of Radiology, Washington University School of Medicine, St. Louis, Missouri

PURDY, JAMES A., Ph.D., Instructor, Radiation Physics, Division of Radiation Oncology, Mallinckrodt Institute of Radiology, Washington University School of Medicine, St. Louis, Missouri

RAWLINSON, JOHN ALAN, M.Sc., Physicist, Ontario Cancer Institute, Toronto, Canada

ROBBINS, ROBERT H., M.D., Professor and Chairman, Department of Radiation Therapy and Nuclear Medicine, Temple University School of Medicine, Philadelphia, Pennsylvania

SCHUMACHER, W., M.D., Professor, Department of Radiation Therapy and Nuclear Medicine, Rudolf Virchow Hospital, Berlin, Germany

TAPLEY, NORAH duV., M.D., Professor of Radiotherapy, The University of Texas System Cancer Center, M.D. Anderson Hospital and Tumor Institute, Houston, Texas

WATSON, THOMAS A., M.B., Director, Ontario Cancer Foundation, London Clinic, Victoria Hospital, London, Ontario, Canada; Clinical Professor and Head, Department of Therapeutic Radiology, University of Western Ontario, London, Ontario, Canada

Series Preface

The past five years have produced an explosion in the knowledge, techniques, and clinical application of radiology in all of its specialties. New techniques in diagnostic radiology have contributed to a quality of medical care for the patient unparalleled in the United States. Among these techniques are the development and applications in ultrasound, the development and implementation of computed tomography, and many exploratory studies using holographic techniques. The advances in nuclear medicine have allowed for a wider diversity of application of these techniques in clinical medicine and have involved not only major new developments in instrumentation, but also development of newer radiopharmaceuticals.

Advances in radiation therapy have significantly improved the cure rates for cancer. Radiation techniques in the treatment of cancer are now utilized in more than 50% of the patients with the established diagnosis of cancer.

It is the purpose of this series of monographs to bring together the various aspects of radiology and all its specialties so that the physician by continuance of his education and rigid self-discipline may maintain high standards of professional knowledge.

LUTHER W. BRADY, M.D.

Introduction

This book presents the proceedings of a symposium held at Thomas Jefferson University in May, 1975. The objective of the conference was to look at the clinical usefulness of very high energy beams and to find out what they can offer over what one might call standard megavoltage in the 2- to 6-MeV range. In a way, the situation is similar to that just before the general introduction of cobalt-60 and 2-MeV units twenty-five years ago. Then radiation therapists fell into three groups: those who knew that supervoltage would be advantageous, those who were not so sure, and those who were sure that it would make no difference. The first group of radiation therapists, of course, already had or were about to have the new machine. The second group did not know whether they could get the equipment in reasonable time. And the third group was pretty sure that they would not be able to get it. Everybody agreed at that time on the physical advantages, but whether a clinical advantage existed was questioned. Yet once the equipment was in use, it became clear that there were clinical advantages. One prominent example was that the scatter component of the beam was reduced sufficiently to allow us to shape our field so that we could decrease the dose of radiation to the transit or normal tissue compartment for an equivalent target-volume dose. We could now choose our target volume in accordance with anatomic and biologic considerations of the tumor to be treated and without the constraint of the field geometry imposed by orthovoltage machines, which allow a limited choice of circular, square, or rectangular fields. Thus, a physical characteristic permitted an important clinical advance.

Will the step beyond standard supervoltage equipment give us comparable clinical advances? How large a step should it be? Is there an ideal energy for photon beams beyond which we are unlikely to get an advantage? Are electrons in the 30- to 45-MeV range of sufficient value in clinical radiation therapy to justify the cost of the equipment? Will a very high energy beam allow us to treat patients with tumor types that we have not been able to treat successfully heretofore? Will we be able to treat patients more aggressively without risking increased normal tissue damage? These are some of the questions that come to mind.

The book is divided into three parts: the first deals with the need for high-energy photons and electrons, the second with current experience with these beams; and the third with, predictions of possible future use of the beams.

We wish to thank Lucinda McC. Pitcairn, for her careful editing and preparation of the manuscripts for the book, and Carolyn Ratliff, who, with great patience and skill, set the type composition for the entire book.

Simon Kramer, M.D.
Nagalingam Suntharalingam, Ph.D.
George F. Zinninger, M.D.

Contents

SECTION I. THE NEED FOR HIGH-ENERGY BEAMS IN RADIATION THERAPY

Desirable Characteristics of High-Energy Photons and Electrons 3
  Harold E. Johns and John A. Rawlinson

Value of and Need for 25- to 35-MV Photon Beams 17
  Gilbert H. Fletcher

Biological Considerations in the Use of High-Energy Beams 37
  Eric J. Hall

The Value of and Need for High-Energy Electrons 51
  John H. C. Ho

Discussion 77

SECTION II. CURRENT EXPERIENCE WITH HIGH-ENERGY BEAMS

Current Treatment-Planning Practice with Photons 85
  John S. Laughlin

Clinical Experience with High-Energy Photons 101
  Thomas A. Watson

Current Techniques in the Use of Nonstandard Fractionation 115
  Jean Dutreix

Dosimetry Considerations of Electron Beams 129
  Peter R. Almond

Clinical Experience with Electron Beams	169
Norah duV. Tapley	
Discussion	197
Problems of High-Energy X-Ray Beam Dosimetry	203
Andrée Dutreix	
High-Energy X-Ray Beams in the Management of Head and Neck and Pelvic Cancers	215
Carlos A. Perez, James A. Purdy, Alvin Korba and William E. Powers	
Electron-Beam Dose Distribution in Inhomogeneous Media	243
W. Pohlit and K.H. Manegold	
The Use of High-Energy Electrons in the Treatment of Inoperable Lung and Bronchogenic Carcinoma	255
W. Schumacher	
Telecentric Rotation with Electron Beams	285
Dieter Fehrentz	
Discussion	301

SECTION III. NEW HORIZONS FOR HIGH-ENERGY BEAMS

The Impact of Radiosensitizers of Hypoxic Cells	309
Jack F. Fowler and G. E. Adams	
A Biophysical Model for Radiation Therapy	321
W. Pohlit	
Optimization of Energy and Equipment	333
Harold E. Johns	
Future Uses for High-Energy, Low-Let Radiation	347
William E. Powers	
Discussion	357
Index	361

High-Energy Photons and Electrons:
Clinical Applications in Cancer Management

Section I

The Need for High-Energy Beams in Radiation Therapy

Desirable Characteristics of High-Energy Photons and Electrons

Harold Elford Johns, Ph.D.,
Professor and Head,
Department of Medical Biophysics,
Ontario Cancer Institute,
Toronto, Ontario, Canada

John Alan Rawlinson, M.Sc.,
Physicist,
Ontario Cancer Institute,
Toronto, Ontario, Canada

In 1958 we installed a 22 MV Allis-Chalmers betatron at the Ontario Cancer Institute. At that time the staff at the Institute started a randomized study to compare the relative merits of cobalt-60 and 22-MV betatron radiation in the treatment of stages IIB and III cancer of the cervix. In 1968 Dr. W.E. Allt reviewed the results of this study (Allt, 1969), reporting that those patients treated with the betatron had a better survival than those treated with ^{60}Co. Figure 1 shows up-to-date survival data for stage III cancer of the cervix from that study.

OPTIMUM DEPTH-DOSE PATTERN FOR HIGH-ENERGY PHOTONS

In Figure 2 we show a schematic representation of the dose distribution which would be optimum in treating a tumor of thickness (d_2-d_1): all the energy in the beam is concentrated in the tumor and none is deposited anywhere else. Such a distribution can never be achieved but may be approached with a high-energy beam of photons, as shown in Figure 2.

DESIRABLE CHARACTERISTICS

Some of the desirable characteristics of such a beam are shown in Figure 2. We require a machine that is *reliable*, i.e., it produces radiation whenever the machine is turned on. It should be able to produce a preselected dose in a *precise* manner. It should yield a *uniform dose* over the area of the beam. Furthermore, the *skin dose* should be as small as possible and preferably zero. It should have a minimal *penumbra* and contain negligible contamination in the form of *neutrons*.

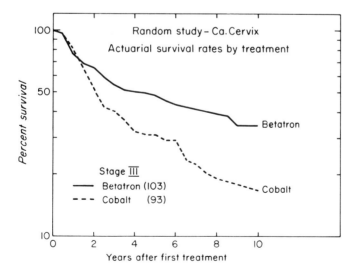

Figure 1. Unpublished survival data for stage III cancer of the cervix based on an update of Allt's earlier work (Allt, 1969).

High Energy Photons

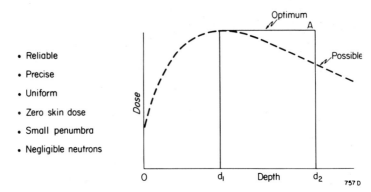

- Reliable
- Precise
- Uniform
- Zero skin dose
- Small penumbra
- Negligible neutrons

Figure 2. Optimum depth-dose distribution and possible distribution for high-energy photons. The tumor is assumed to extend from d_1 to d_2.

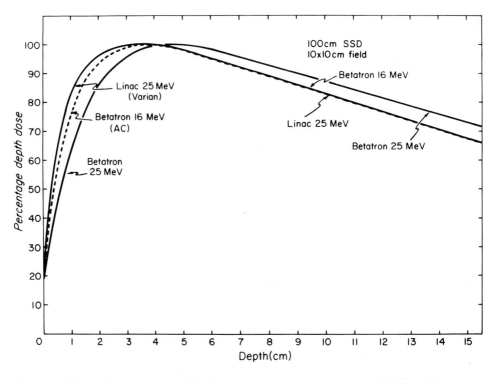

Figure 3. Depth-dose from our Allis-Chalmers betatron operating at 25 MV and from our Varian linac operating at 25 MV, 10 cm × 10 cm at 100 cm. Also shown is the depth dose for the betatron operating at 16 MV (Rawlinson and Johns, 1973).

DEPTH-DOSE FROM LINACS AND BETATRONS

In Figure 3 we present depth-dose data for our betatron operating at 25 MV and compare it to the distribution from our "Clinac 35," operating at 25 MV.

It is clear that the betatron beam is more penetrating, has less skin dose and delivers its maximum dose at a greater depth than the linac. We then asked ourselves what betatron energy would produce the same dose pattern as the linac. Figure 3 shows that our betatron operating at 16 MV is essentially equivalent to the 25-MV linac beam from a dose distribution point of view.

In order to explain this difference between the two machines, one must examine the way in which the x-rays are produced in the betatron and in the linac. In the betatron (Fig. 4) electrons travel in an expanding circular orbit until the electron beam hits a very small tungsten target. If an electron collides with an atom in the target and loses some energy, it will spiral in towards the center of the doughnut, because of the magnetic field of the betatron, and most of the energy loss will appear as an x-ray photon. The x-ray beam which emerges from such a system is called thin-target radiation. If the electron goes through the target without losing any energy, it will continue around the circular orbit and probably interact with the target on the second pass. *In the betatron every x-ray photon produced was produced by an electron which had the full energy E_{max} of the electron beam.*

In contrast, we see what happens in the linac (Fig. 4). The electron beam is now incident on a thick target. This thick target is essential to prevent electrons from passing through it and so hitting the patient. *The photons are now produced by an electron beam which may have any energy from E_{max} down to zero.*

Another difference between these two machines is the way in which the flattening filter is designed. In the betatron the flattening filter is usually made of some material with a low atomic number, such as aluminum (Johns et al., 1949; Johns, 1950), whereas in the linac, in order to save space, the compensating filter has been made of either heavy metal

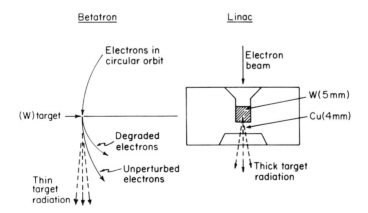

Figure 4. A schematic diagram to illustrate the difference between thin-target radiation produced in the betatron and thick target radiation produced in the linac.

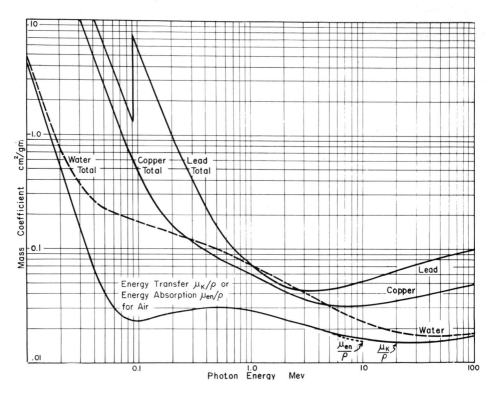

Figure 5. Attenuation coefficients for photons as a function of energy. (Reprinted, with permission, from Physics of Radiology, 1969)

or lead. The material used in the flattening filter has an effect on the photon beam, and this can be appreciated by an examination of Figure 5, which shows the attenuation coefficients for photons as a function of energy for lead, copper and materials with a low atomic number, such as water. At energy levels above 3 MeV, the attenuation coefficient in Pb and Cu increases with increasing energy. If one uses lead as a flattening filter, one will preferentially select against both high-energy photons and very low-energy ones in favor of photons at 3-4 MeV. If sufficient lead filtration is used, a beam with high- and low-energy photons will be degraded until all the photons have an energy of about 3 or 4 MeV, which is the minimum in the curve shown in Figure 5. The use of aluminum, on the other hand, will cause a preferential selection against low-energy photons only.

Thus, we have two major differences between the linac as made by Varian and the betatron as made by Allis-Chalmers. The betatron uses thin-target radiation and an aluminum flattening filter, whereas the linac uses thick-target radiation and a lead flattening filter. This whole problem is quite complex; to find the optimum design, we embarked on a series of experiments with our linac. The way this was done is shown in Fig. 6.

In the "Clinac 35" the electron beam is bent through 57°; then, after passing through two quadruple magnets, it is normally bent through 90° to hit a tungsten target and produce x-rays through port No. 3. By turning off the 90° magnet, it is possible to have the

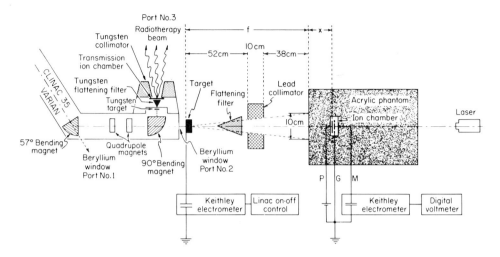

Figure 6. Experimental arrangement to study the effects of target and filter design on the depth-dose distribution (Podgorsak et al., 1974).

electrons emerge through a thin window into the air with negligible loss of energy. These electrons can then be directed on to targets of various design to produce an x-ray beam which is passed through flattening filters made of different materials.

In a series of experiments (Podgorsak et al., 1974), we have investigated the effect of target design, filtering material and electron energy on the depth-dose. These experiments showed that both the filter material and the target design affect the depth-dose. After trying many combinations, we showed that a thick target of aluminum and a flattening filter of aluminum gave a near optimum beam. The Atomic Energy of Canada, Ltd. group in Ottawa have shown (Sherman et al., 1974) that a slight further improvement can be achieved with thin- rather than thick-target radiation from aluminum, but they have not demonstrated this advantage in a therapy situation. Some of our results are shown in Figure 7, in which we compare the 25 MV betatron with the Varian linac operating at 25 MV, in which the target and flattening filter have both been replaced by one made of aluminum. A linac altered in this way will give an excellent dose distribution, and because of the linac's very high yield, it should be more useful in radiation therapy than the betatron. We have not yet redesigned the head of the Varian machine to include these improvements. It is hoped that the designers of new linacs will take these suggestions seriously.

The choice of an aluminum target might appear a bit surprising, since it is generally assumed that the x-ray yield depends on Z and that an aluminum target would give a much lower yield than a tungsten or lead one. That this is not true is illustrated in Figure 8, where we show the ion chamber response as a function of depth for the same electron current into targets of various thickness. The yield at the maximum is essentially independent of Z. This is true provided we look at the yield in the forward direction. If one measures the yield at larger angles than a few degrees, then the yield is very much a function of Z. For example, at $30°$ the beam from Pb is twice the intensity of one produced in Al. The integrated yield over all angles depends upon Z to the first power.

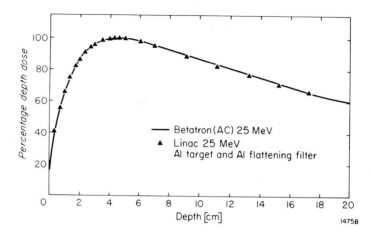

Figure 7. Depth-dose from 25 MV betatron using thin-target x-radiation and an aluminum flattening filter, compared to linac radiation produced in a thick aluminum target and an aluminum flattening filter.

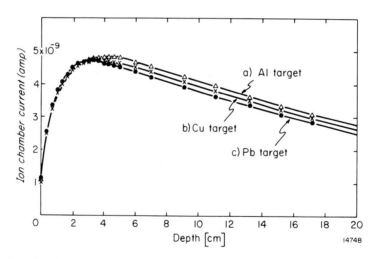

Figure 8. Ion chamber response as a function of depth for beams produced in targets of aluminum, copper and lead (Podgorsak et al., 1974).

In the last 10 years there have been a number of linacs produced with energies in the 4-18 MV range. To see if the conclusions given above also applied at these lower energies, we used the arrangement shown in Figure 6 to produce lower-energy photon beams (Podgorsak et al., 1975). Some of our results are shown in Figure 9, where the 10, 15, 20 and 25 cm depth-doses are plotted as a function of photon energy for various combinations of target material and flattening filter. At 25-30 MV the materials used for the target

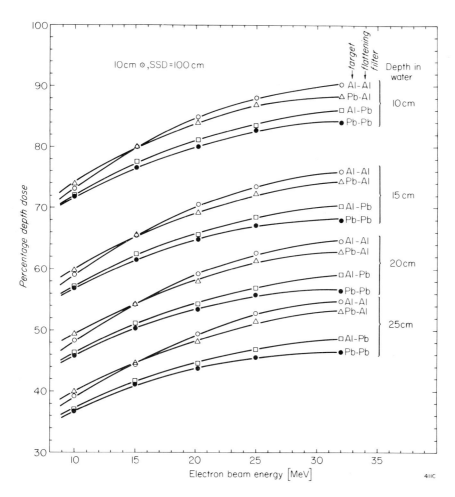

Figure 9. Percentage depth dose at 10, 15, 20 and 25 cm depth for various filter and target combinations as a function of the electron energy used to produce the x-ray beam. Circular field of 10 cm diameter at an SSD of 100 cm (Podgorsak et al., 1975).

and filter have a large effect on the depth-dose. In contrast, at 10 MV, the effect of these parameters is much smaller, and the design of the filter and target are less critical. This graph also shows that very little improvement in the photon depth-dose is achieved with an increase in the electron energy above 25 MV. There seems to be little justification for using linacs in the photon mode with energies above 30 MV.

LARGE FIELDS

With our Varian Clinac 35 we can use fields up to 35 cm × 35 cm at 1 meter, and this large field capability is very useful. However, large fields introduce a number of problems.

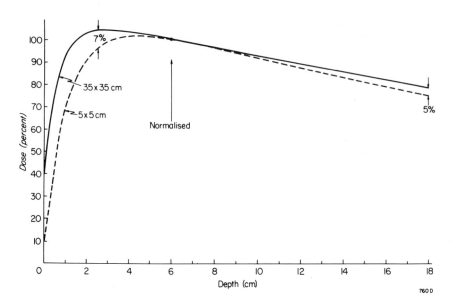

Figure 10. Depth dose for small (5 cm × 5 cm) and large field (35 cm × 35 cm) from our "Clinac 35" operating at 25 MV (Unpublished data).

Figure 10 shows the depth-dose data for a small and a large field. With the large field, the surface dose is larger, and the position of the maximum is shifted some 2 cm nearer the surface. The extra dose from 0 to 4 cm is almost certainly due to low-energy photons, electrons or both. Leung and Johns (in press) have shown that similar contamination occurring in large cobalt fields can be removed by the use of filters. See also Saylor and Quillen (1971) and Rao et al. (1973). We ourselves have investigated the effect of filters of various atomic numbers and thicknesses at various positions in the linac beam. Some results are shown in Table 1. The distribution with the minimum surface dose and deepest peak dose was obtained using an aluminum filter (0.25 in.) above the jaws and a Pb filter (0.005 in.) between the jaws. Our experiments, which have been reported elsewhere (Galbraith et al., 1975), suggest, but don't prove, that the contamination is mainly electrons from the flattening filter. However, we only partially understand the problem, and further work on this topic is required.

The table also shows that one cannot predict the depth-dose from electron energy and field size alone, because of the unknown contribution of scatter which depends on the design of the head.

ELECTRONS

When one uses high-energy electrons in radiotherapy, one is usually interested in treating a volume extending from near the surface depth d_1 to a depth d_2, as illustrated in Figure 11, and one would ideally like the optimum dose distribution shown in the figure. If the tumor extends to the skin surface, then the addition of a layer of bolus can bring the

TABLE 1. EFFECT OF FILTERS ON DEPTH DOSE FOR A NUMBER OF CONFIGURATIONS: 25 MV LINAC BEAM (30 CM X 30 CM)[a]

Depth (cm)	Depth Dose (%)					
	Open Beam	Al below jaws	Pb below jaws	Pb above jaws	Al above jaws	Al above and Pb between jaws
0	37.7	40.1	35.4	29.7	32.3	26.8
1.4	93.9	93.6	91.7	89.5		87.2
2.0	98.6	98.8	97.2	95.6	96.5	94.6
2.7	100.0	99.6	100.0	97.8	99.0	98.5
3.4	100.0	100.0	100.0	100.0	100.0	99.5
4.0	97.9	99.6	99.6	100.0	99.5	100.0
5.0	96.4	98.5	98.4	98.7	98.2	98.7
9.3	89.5	92.0	92.1	91.4	91.4	91.8
18.0	75.6	77.4	77.4	77.3	77.2	77.8

[a]Galbraith and Rawlinson, unpublished data. Al filter—0.25 in. Pb filter—0.005 in.

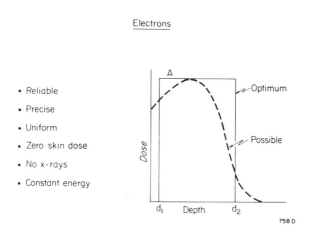

Figure 11. *Optimum and possible depth-dose from electrons.*

high-dose region at B to the surface of the skin. In practice, one can only approach this optimum distribution.

As before, we want *reliability*, *precision* in dose control, *uniform dose distribution*, *constant energy* during a given treatment, and *minimum x-ray contamination*. For electrons one also wants a variable selection of fields from small to very large ones. When large fields are used, there is a problem in achieving a uniform dose over the large area. A number of methods have been used to achieve the uniform distribution. The most common one is to pass the electron beam through a Pb foil to scatter it into a wider beam. The thicker the foil, the more the beam is scattered, but also, of course, the x-ray

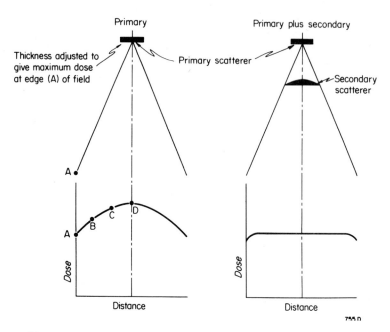

Figure 12. Use of two scattering foils to obtain a uniform electron distribution.

contamination is increased. More sophisticated techniques are also available—these include scanning the electron beam over the desired area. We have had no experience with such electron scanners; however, we have found an excellent way to obtain a uniform beam using two foils as shown in Figure 12.

Before describing these foils it is instructive to examine the scattering problem in detail. We have found that the scattering by a foil gives rise to a Gaussian-like distribution, as illustrated in Figure 13. The parameters in this equation can be determined experimentally.

Suppose we wish to obtain a uniform dose distribution out to A in Figure 12. We determine the thickness of the primary foil required to give the maximum dose at A for a given electron current. There is obviously one thickness t which will make the dose at A largest— since too thin a foil will not scatter enough to A, while too thick a foil will scatter too much outside A. Having selected this thickness, we then find the doses at points such as B, C and D are too large. Theses doses can be reduced by adding a compensating scatterer (Fig. 12). Its design can be carried out with the aid of a computer as follows: We first calculate the dose profile $ABCD$. The doses at BC and D are too large, and we then estimate, using the formula of Figure 13, how much lead should be placed in the compensating scatterer at each radius to reduce the doses to BC and D to the value at A. We then recalculate the doses at B, C and D, taking into account the scatter. Such a calculation will show that the doses at B, C and D are still too large. Therefore, we estimate a correction to the thickness of the scattering filter and redo the calculation. After some ten iterations the dose profile can be made acceptably flat,

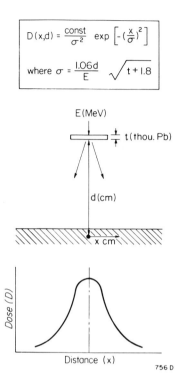

Figure 13. Gaussian distribution of electrons scattered from a Pb scattering foil.

as illustrated in Figure 12. For details of this calculation, see Rawlinson and Galbraith (in press). We believe designing the filters in this way will lead to minimum x-ray contamination. We have found that to flatten a 20 MV beam over a 35-cm diameter field at 1 meter, the primary filter should be about 0.12 in. of Pb and the flattening filter should vary from 0 to 0.012 in. of Pb at the center. The filter can be made with 6 concentric discs of lead, each 0.002 in. thick.

SUMMARY

In this paper we have suggested a number of ways in which present-day linacs can be improved in order to give a more desirable radiation beam. We have shown that the depth-dose achieved is a function of many parameters other than the electron energy and the area of the beam. We have also discussed one way in which contamination of a high-energy beam, when large fields are used, can be partially controlled. We have shown how two scattering foils can be used together to give a uniform electron distribution for electron therapy.

ACKNOWLEDGMENTS

The authors take pleasure in acknowledging the financial assistance of the National Cancer Institute of Canada and the Medical Research Council of Canada. We are indebted to Dr. W.E.C. Allt for making his unpublished survival data for cancer of the cervix available, and to Dr. Ervin Podgorsak and Mr. Duncan Galbraith for their help in taking many careful measurements.

REFERENCES

Allt, W.E. 1969. Supervoltage radiation treatment in advanced cancer of the uterine cervix: A preliminary report. *J Can Med Assoc* 100:792–797.

Galbraith, D.M., J.A. Rawlinson & H.E. Johns. 1975. The effect of collimation design on the depth dose for high energy x-ray linear accerlators (Abstract). *Med Phys* 2:166.

Johns, H.E. 1950. The betatron in cancer therapy. *Nucleonics* 7:76–83.

Johns, H.E., E.K. Darby, R.N. Haslam, L. Katz & E.L. Harrington. 1949. Depth dose data and isodose distributions for radiation for a 22 MeV betatron. *Amer J Roentgenol* 62: 257–268.

Leung, P.M.K. & H.E. Johns. Optimum filter for cobalt-60 therapy units, in press.

Physics of Radiology, 3rd Edition. Charles C. Thomas, Springfield. 1969.

Podgorsak, E.B., J.A. Rawlinson, M.I. Glavinovic & H.E. Johns. 1974. Design of x-ray targets for high energy linear accelerators in radiotherapy. *Amer J Roentgenol* 121: 873–882.

Podgorsak, E.B., J.A. Rawlinson & H.E. Johns. 1975. X-ray depth doses from linear accelerators in the energy range from 10 to 32 MeV. *Amer J Roentgenol* 123:182–191.

Rao, P.S., K. Pillai & E.C. Gregg. 1973. Effect of shadow trays on surface dose and build-up for megavoltage radiation. *Amer J Roentgenol* 117:168–174.

Rawlinson, J.A. & D.M. Galbraith. *Med Phys*, in press.

Rawlinson, J.A. & H.E. Johns. 1973. Percentage depth dose for high energy x-ray beams in radiotherapy. *Amer J Roentgenol* 118:919–922.

Saylor, W.L. & R.M. Quillin. 1971. Methods for the enhancement of skin sparing in cobalt-60 teletherapy. *Amer J Roentgenol* 111:174–179.

Sherman, N.K., K.H. Lokan, R.M. Hutcheon, L.W. Funk, W.R. Brown & P. Brown. 1974. Bremsstrahlung radiators and beam filters for 25-MeV cancer therapy. *Med Phys* 1:185–192.

Value of and Need for 25- to 35-MV Photon Beams

Gilbert H. Fletcher, M.D.,
Professor and Head,
Department of Radiotherapy,
The University of Texas System Cancer Center,
M.D. Anderson Hospital and Tumor Institute,
Houston, Texas

In July, 1954, a 22-MeV Allis-Chalmers betatron was put into clinical use at M.D. Anderson Hospital. In 1963, an 18-MeV Siemens betatron with photon and electron beams was acquired. In 1971, a 25-MeV linear accelerator was installed, and at the same time the initial Allis-Chalmers unit was replaced by a new model, which operates at 25 MeV. In the period from July, 1954 through 1974, approximately 10,000 patients have been treated with 22-25 MeV and about 1,000 with 18-MeV photons. Patients with cancer of the cervix constitute the largest group treated with 22-25 Mev.

There is little difference in depth-dose and build-up between 22 and 25 MeV. The build-up with 18 MeV is not as deep and the depth-dose is a little less, so this beam is used in treatment of head and neck cancers. The 25-MeV beam is adequate for the majority of the deeply located tumors.

TABLE 1. PHOTON BEAM DOSIMETRY (Patient Diameter, 25 cm)

Depth Below Surface (cm)	^{60}Co	4 MV	6 MV	8 MV	10 MV	25 MV	35 MV
1	1.20	1.14	1.11	1.02	1.00	0.78	
2	1.17	1.14	1.11	1.08	1.04	0.92	0.85
3	1.15	1.12	1.09	1.08	1.06	0.99	0.93
4	1.12	1.10	1.08	1.06	1.08	1.02	0.98
5	1.10	1.07	1.06	1.05	1.07	1.03	1.00
6	1.08	1.06	1.04	1.04	1.06	1.03	1.01
7	1.06	1.04	1.03	1.03	1.04	1.02	1.01
8	1.04	1.02	1.02	1.02	1.03	1.01	1.01
9	1.02	1.01	1.01	1.01	1.02	1.01	1.00
10	1.01	1.00	1.01	1.01	1.01	1.00	1.00
11	1.00	1.00	1.00	1.00	1.01	1.00	1.00
12	1.00	1.00	1.00	1.00	1.00	1.00	1.00
12.5	1.00	1.00	1.00	1.00	1.00	1.00	1.00

Central-axis depth-dose for parallel-opposing fields (15 X 15 cm), 100-cm treatment distance. All values are relative to mid-diameter dose. (Reprinted, with permission, from Levitt, 1973)

This investigation was supported in part by Public Health Service Research Grants CA-06294, CA-05654, and CA-05099 from the National Cancer Institute.

CLINICAL BENEFITS

The clinical benefits of 25-MeV photons derive essentially from the physical characteristics, which are:

1. A considerable increase in depth-dose over the 3- to 6-MeV level; and
2. A significant increase in the depth of maximum build-up.

Figure 1 compares the dose to the subcutaneous tissues with the midline dose for 4 MeV and 25 MeV in a 15-cm thick part, (i.e., nasopharynx and oropharynx) and in a 25-cm thick part (i.e., average AP diameter for a male or female pelvis).

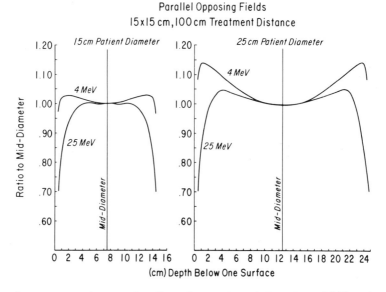

Figure 1. Comparison of ratio of midline dose to lateral dose for 4 MeV and 25 MeV at 100 cm TSD for two widths of separation. (Reprinted, with pemission, from Fletcher, 1973)

Table 1 shows that, with an AP diameter of 25 cm, even with a 10 MeV beam, the midline dose is lower than the dose in the subcutaneous tissues. The table also shows that 35 MeV is slightly better than 25 MeV for particularly thick patients. Because of the considerable depth-dose, in addition to the favorable dose distribution with parallel-opposing portals, lateral portals can be used effectively with 25 or 35 MeV to treat pelvic tumors.

As well as providing the obviously superior volume distribution, which results, in considerable skin-sparing with lesser doses to the subcutaneous tissues, the use of portals that are always anterior, posterior, and lateral, assures a safe anatomical coverage and an easy daily duplication of the treatment set-ups. Only with AP, PA, and lateral portals can

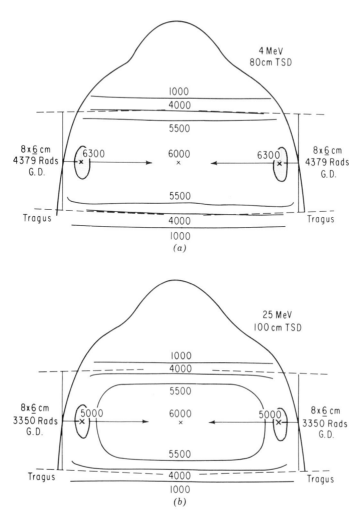

Figure 2. Dose distribution with 4 MeV (Fig. 2a) and 25 MeV (Fig. 2b) for irradiation of nasopharyngeal tumor using parallel-opposed portals entirely. With 25 MeV for a central dose of 6000 rads, the dose to the temporomandibular joint is reduced from 6300 to 5000 rads.

structures be clearly identified on the verification films. The structures on oblique films are so distorted that there can be no meaningful identification. Furthermore, with oblique portals there is an uncertain anatomical coverage. For example, initially at the M.D. Anderson Hospital, patients with cancers of the urinary bladder were treated by three fields—one AP and two posterior-oblique portals. Random checking of treatment portals by verification films during treatment with the bladder filled with air showed that the anatomical coverage with the posterior-oblique fields was marginal, unless very wide portals were used. With the patient prone, a slight difference in position exceeds the marginal anatomical coverage of the posterior oblique portals.

OPTIMIZATION OF VOLUME DISTRIBUTION WITH THE 25-MEV BEAM

Used alone, the 25-MeV photon beam produces an optimal volume distribution in a number of clinical situations. Rotation therapy is not a substitute for the "box" technique from the standpoint of integral dose. Rotation is, in fact, a very inefficient way to deliver irradiation. The acute reactions will be excessive when the target volume is greater than 7-8 cm in diameter. If the target volume is the whole pelvis, i.e., a minimum of 14 cm in diameter, the patient experiences considerable diarrhea because a large volume of bowel receives 50% or more of the tumor dose given at the periphery of the target volume.

The 25-MeV beam can be used for optimization of treatment in combination with 3- to 6-MeV beams and also with electrons. A combination of 25 MeV photons, 3- to 6-MeV photons, and electrons may be used in the same patient, depending upon the anatomical areas involved.

Examples are given below of indications for the use of 25-MeV photons alone or combined with 3- to 6-MeV photons, 7- to 19-MeV electrons, or both, for various anatomical sites.

HEAD AND NECK

For tumors of the nasopharynx, antral portals should not be used if there is invasion of the base of the skull, cranial nerves, or both, or extension along the pharyngeal walls, since only parallel-opposed portals can cover these areas. A minimal tumor dose of 6500 or even 7000 rads is needed, resulting, with 3- to 6-MeV parallel-opposed portals, in doses of 7000 and 7500 rads to the temporomandibular joint and the masseter muscle (Fig. 2a). Severe fibrosis of these anatomical structures will ensue. With 25 MeV, the dose to the lateral tissues is significantly diminished (Fig. 2b). In centrally located tumors, like those of the uvula, parallel-opposed portals with 25 MeV produce an elegant volume distribution, with an adequate dose to clinically negative subdigastric nodes (Fig. 3).

One of the major late sequelae of irradiation in oral cavity and oropharyngeal tumors is dryness, which can be moderately discomforting or very crippling, since it produces loss of appetite. Patients may have to carry a water bottle to repeatedly moisten the mucosa. A single homolateral portal with 3 to 6 MeV for unilateralized tumors, like those of the tonsillar area, results in too high a dose to the subcutaneous tissues. With 25 MeV there is a gradient of dose through the oropharyngeal mucosa which results in a clear difference in the intensity of the mucositis between the involved and contralateral sides of the oropharynx (Fig. 4). The lesser mucosal reaction translates later into better moisture of at least half of the oropharynx. When the nodes in the neck are clinically negative, there is enough thickness of the skin and subcutaneous tissues for the ipsilateral nodes to be included in the area of at least the 80% isodose curve. The opposite subdigastric nodes will still receive at least 5000 rads dose with a tumor dose to the primary of 7000 rads at 3-3.5 cm. With clinically positive nodes, more so if large, tumor can be only a millimeter or two under the skin and the 3- to 6-MeV beam must be used to deliver at least 5000 rads.

For very lateralized lesions, like those of the retromolar trigone and the foot of the anterior faucial pillar, if the nodes are clinically negative, only the ipsilateral subdigastric lymphatics need to be irradiated, since the incidence of contralateral metastases is small

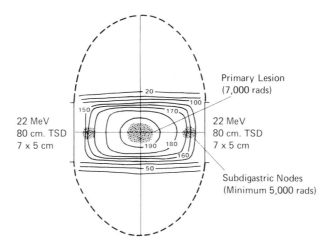

Figure 3. Dose distribution with parallel-opposed portals with 22 MeV for a squamous cell carcinoma of the uvula. (Modified from Fletcher, 1973)

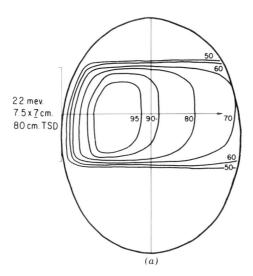

(a)

Figure 4. Male, age 58, seen April 4, 1955, with a lesion, approximately 3 cm in diameter, of the tonsillar pillar invading the adjacent base of the tongue. There were no palpable nodes. Histology: Squamous cell carcinoma, Grade III. (a) Single, homolateral, portal volume distribution with 22 MeV beam. This is a gradient approximately 10-15% across the oropharynx. (b) A tumor dose of 5600 rads (taking the 95% curve) was given in four weeks with 22 MeV. A confluent mucositis developed on the right retromolar trigone-anterior faucial pillar. The area of the soft palate on each side of the midline was at approximately the 90% level and exhibited a studded mucositis. Only marked redness appeared on the opposite faucial pillar and retromolar trigone which were within the 80-85% curve. A difference in dose of 10-15% from one side of the palatine arch to the other results in a gradient of reaction ranging from a confluent mucositis to studded mucositis and an angry red mucosa. (Reprinted, with permission, from Fletcher, 1956, 1973)

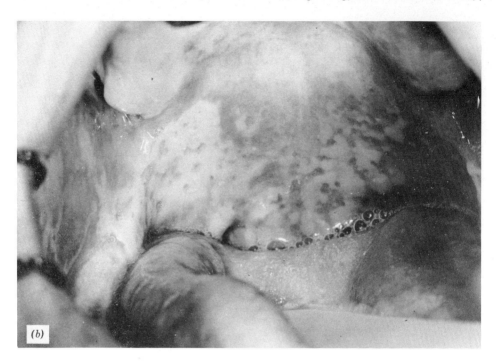

Fig. 4 (Continued)

with clinically negative, ipsilateral, subdigastric nodes. A combination of half 25-MeV photons and 19-MeV electrons or 18-MeV photons and electrons produces an optimal volume distribution (Fig. 5). There is no acute mucositis on the contralateral side of the oropharynx, and the dose to the opposite parotid is only the percentage which is received from the photon beam, since there is zero contribution from the electron beam. These patients have a normally moistened mucosa.

The failure rate in lesions of the posterior or lateral pharyngeal walls, even with doses of 7000 rads, is the highest of all lesions of the oropharynx. Characteristically these lesions have "skip areas" with submucosal spread of disease. Overly generous coverage must be used to include the parapharyngeal lymphatics to the base of the skull (Fig. 6). Whenever the nodes are clinically negative, the 25-MeV beam with parallel, opposed portals is recommended. After irradiation with 25 MeV, the tissues are excellent, and if there is a recurrence, there is no problem of wound healing after the required, extensive, surgical procedure.

Tumors of the nasal cavity and ethmoids can be optimally treated by a combination of 3-6 MeV and 25 MeV photons, using equal given doses. If the wall of the orbit is not involved, the patient gazes straight ahead, and the lateral edge of the field just glances the cornea. Depending upon the depth selected for the minimum tumor dose, the combination of half and half 25 MeV and 3-6 MeV photons through a single anterior portal may be adequate (4-6 cm depth) or lateral, wedged portals can be added (7-8 cm depth; Fig. 7). If the wall of the orbit is destroyed, a wider anterior portal must be used. The eyes are turned outward, and one must always use lateral wedge portals to avoid an excessive dose to the macula and optic nerve (Fig. 8; Shukovsky and Fletcher, 1972).

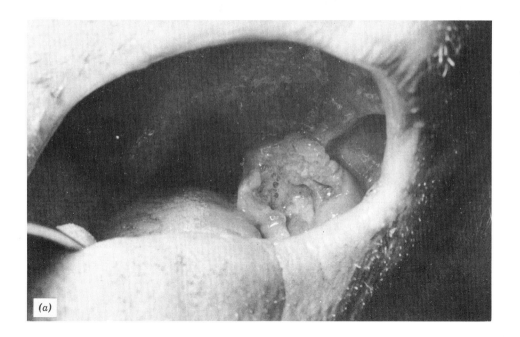

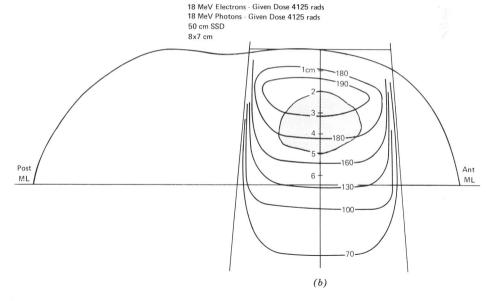

Figure 5. Male, age 62, first seen at M.D. Anderson Hospital in May 1966 with a 3-cm exophytic ulcerated lesion of the left retromolar trigone, extending into the base of the anterior tonsillar pillar, glossogingival fold, and postero-lateral tongue border. There was no adenopathy in the neck. Biopsy showed squamous carcinoma, Grade II. The patient was treated with combined 18 MeV photons and electrons using equal given doses (4275 rads) with each. The subdigastric nodes were included in the large treatment field and received at least 6000 rads. The calculated tumor dose at 5 cm was 7000 rads in 6½ weeks. The final 1000 rads was given through a 5-cm circle. Confluent mucositis developed on the involved side and none on the opposite side. The patient has remained free of disease and was doing well in July, 1974. (a) Lesion before treatment. (b) Volume distribution of combined beams. (c) Mucositis. (Figs. 5a and b, reprinted, with permission, from Tapley, 1976)

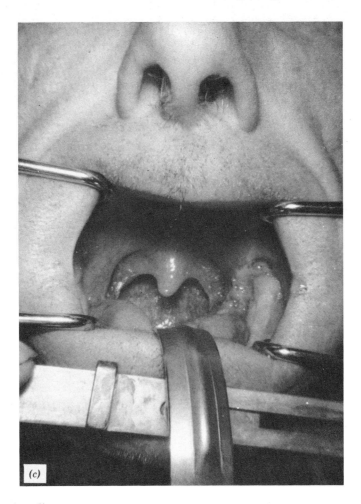

Fig. 5 (Continued)

There are clinical situations where, in addition to the lower neck, the mediastinum must be irradiated. The lower neck portal is extended to cover the mediastinum (Fig. 9). In addition to 5000 rads given dose, the mediastinal part of the field is irradiated with 25 MeV to give a minimum tumor dose of 5000 rads at a depth of 6 or 7 cm. With this technique, the spinal cord is not over-irradiated. With some tumors of the thyroid, irradiation is given to the entire neck and the upper mediastinum. A given dose of 5000 rads is delivered with 3-6 MeV, with an additional 1000 rads to the neck with 9-MeV electrons. In addition, 1000 rads are given to the mediastinal portion of the field with 25 MeV (Fig. 10). In tumors of the cervical esophagus and of the thoracic inlet, oblique portals with compensators are very uncertain, and positioning them every day is time-consuming. Using 3-6 MeV for the anterior portal (including the lower neck) to provide a given dose of 6000 rads and 25 MeV for the posterior portal (mediastinum only), 6000 rads can be given to the tumor without exceeding 5000 rads to the spinal cord (Fig. 11).

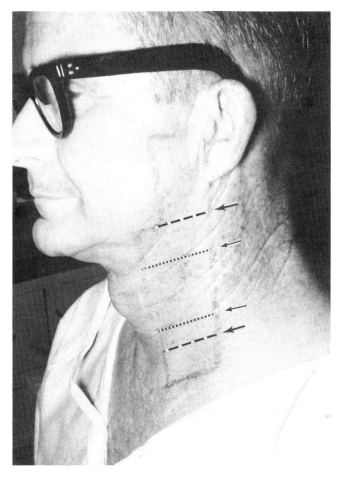

Figure 6. Male, age 54, with an exophytic tumor arising on the posterior hypopharyngeal wall extending to the left pyriform sinus. Biopsy showed squamous cell carcinoma; the biopsy from the anterior aspect of the left pyriform sinus showed in situ carcinoma. From August 22, 1969, to October 13, 1969, 5000 rads were delivered with 18 MeV-photons in 5 weeks, 5 days a week, through parallel-opposed portals from the base of t' skull to close to the sternum. The extension to the base of the skull was to treat the par, arngeal lymphatics, and the lower extension was generous enough not to miss nondetectable extension into the cervical esophagus. At 5000 rads, the portals were reduced (outer dotted lines) to include segment of the pharynx and esophagus above and below the initially observed lesion; from 6000-7000 rads the portals were further reduced (inner dotted lines). During treatment, moderate edema of the arytenoid developed. Only a patchy mucositis was seen at the level of the hypopharynx. The patient had no evidence of disease in February 1975. The tissues of the neck were soft. One could not detect that the patient had received heavy irradiation. (Reprinted, with permission, from Fletcher, 1973)

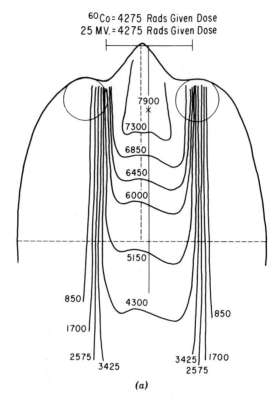

Figure 7. Mixed modality irradiation of ethmoid sinuses and nasal cavity. There is no destruction of the medial wall of the orbit. The eyes are positioned straight forward. (a) Isodose distribution with equal loading, single-entry ^{60}Co and 25-MeV photons. There is an unfavorable ratio of dose between the tumor dose delivered to the posterior ethmoid cells and the dose delivered to the lateral retina, optic nerve, and chiasm. This latter dose exceeds the minimum tumor dose by approximately 12%. (b) A more favorable ratio of doses with relative homogeneity is achieved by adding posterior, lateral wedges with ^{60}Co, thus eliminating the excessive dose to the retina and optic nerves. The ratio of given doses is:

    Anterior ^{60}Co portal 2
    Anterior 25 MeV portal 1
  Right and left lateral ^{60}Co 60° wedges 0.33

(Reprinted, with permission, from Levitt, 1973)

TUMORS OF THE THORAX

Tumors of the esophagus are best treated posteriorly with 25 MeV and anteriorly with 3-6 MeV. With this scheme, 5000 rads can be given without over-irradiating the spinal cord. Oblique portals for tumors of the upper and middle thirds of the esophagus provide very marginal coverage of the esophagus, because of the lordosis. Furthermore, the mediastinal nodes are not well covered.

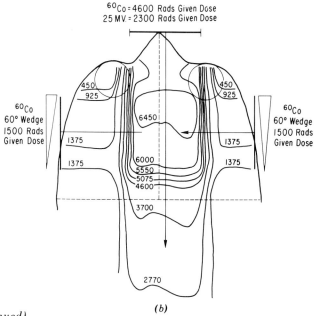

Fig. 7 (Continued)

Bronchogenic carcinoma generally would be treated best anteriorly with 3-6 MeV and posteriorly with 25 MeV (Fig. 12). The anterior portal can then irradiate the supraclavicular nodes, if desired. A given dose of 5000 rads is delivered anteriorly, and the posterior given dose is determined by the dose one wishes to give the tumor.

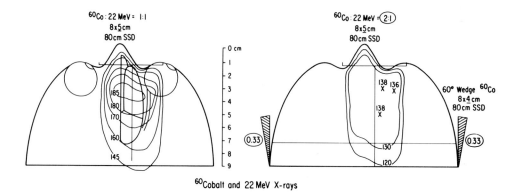

Figure 8. Isodose distribution with half ^{60}Co and half 22-MeV x-rays, with or without lateral wedged portals, for a tumor of the ethmoids which had destroyed the medial wall of the orbit. The eyes are turned laterally to allow a larger part of the orbit to be in the treated volume without irradiating the anterior chamber of the eye. The globes and right optic nerve have been sketched in to show their position within the volume distribution. (Reprinted, with permission, from Shukovsky & Fletcher, 1972)

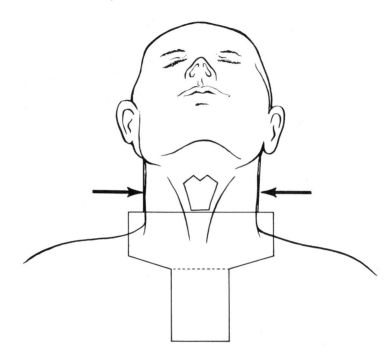

Figure 9. A given dose of 5000 rads is delivered to the lower neck and mediastinal portal. With the 25-MeV beam, additional irradiation is given to the mediastinal part of the portal to bring the dose at 5- to 6-cm depth also to 5000 rads.

PELVIC TUMORS

With 25 MeV, 4000 or even 5000 rads can be given to the pelvis through parallel, opposed portals with absolutely no late subcutaneous fibrosis. Also with 25 MeV, 1000 rads per week are tolerated without diarrhea, whereas only 850-900 rads per week can be given with 3-6 MeV. With 25 MeV, the lateral loops of small bowel receive less than 1000 rads per week, whereas with 3-6 MeV they receive more than 1000 rads per week.

The so-called "box" technique, that is, anterior, posterior, and lateral portals, is optimal for delivery of more than 5000 rads in late-stage tumors of the uterine cervix and all tumors of the urinary bladder and prostate (Fig. 13). At the M.D. Anderson Hospital, few patients with cancers of the uterine cervix or of the endometrium have been treated with ^{60}Co, and, therefore, we have no series to compare survival rates of other treatment methods. At Princess Margaret Hospital in Toronto, the survival rates were clearly better in patients treated with 22 MeV on the Allis-Chalmers betatron than those treated with ^{60}Co (Allt, 1969). Frank Ellis (Personal Communication) stated that while he was still in England, he, as well as his other English colleagues, could not understand how we could give the high doses published for the M.D. Anderson Hospital patient series. After having used an Allis-Chalmers betatron in Milwaukee, he changed his opinion and stated that higher doses could be given with 25 MeV than with 3-6 MeV.

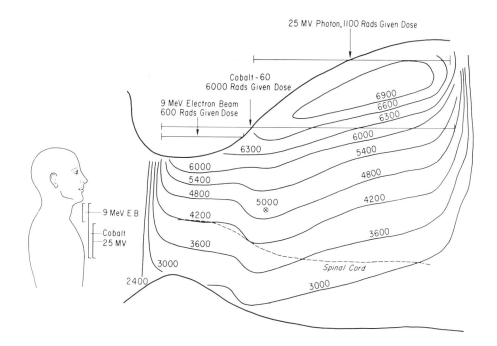

Figure 10. Isodose distribution with anterior ^{60}Co, 25-MeV x-ray and 9-MeV electron-beam portals to cover the lower neck and the upper mediastinum. This technique is selected for anteriorly placed, upper mediastinal lesions, such as thyroid carcinoma with mediastinal involvement. An anterior, 25-MeV portal is added to build up the dose to the anterior mediastinum. The 9-MeV electron beam is used to boost the dose to the anterior neck. The dose to the spinal cord does not exceed 4200 rads. (Reprinted, with permission, from Levitt, 1973)

CLINICAL ADVANTAGES

With 25 MeV, the mucositis is diminished in patients with tumors of the upper respiratory and digestive tracts, and patients treated with large portals for pelvic tumors have a better tolerance. The total absence of late skin changes and subcutaneous fibrosis is the most striking feature. The tissues at 5 or 10 years in head and neck patients treated with parallel, opposed portals with 22 MeV are soft compared with those treated with ^{60}Co. For various reasons, a few patients with cancer of the cervix, or endometrium, or both, were treated with ^{60}Co parallel, opposed portals. Every one of these patients, even with only 4000 or 4500 rads midline doses, later had some subcutaneous induration. When extended portals for cancer of the cervix were initiated, patients were treated on ^{60}Co until the 25-MeV linear accelerator was available. Patients alive several years later had severe subcutaneous changes.

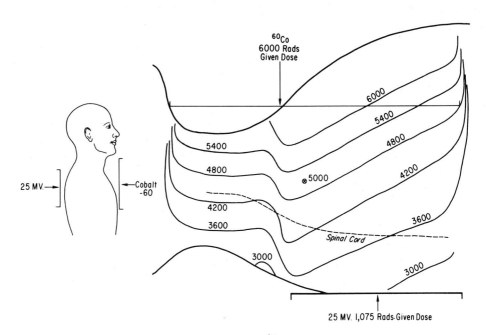

Figure 11. Isodose distribution with an anterior ^{60}Co portal which covers the lower neck and the upper mediastinum. This technique is selected for carcinomas of the upper one third of the esophagus. A posterior, 25-MeV portal is added to build up the dose in the mediastinum, but the lower neck is excluded to avoid overdose to the spinal cord. Sufficient dosage is delivered via the posterior 25-MeV field to provide 5000 rads to the upper mediastinum at mid-diameter. The given dose with ^{60}Co to the lower neck is 6000 rads. (Reprinted, with permission, from Levitt, 1973)

CONCLUSION

The 25-MeV beam has a distinct advantage. It can be used alone or combined with the 3-6 MeV photon beam or the electron beam, or both, to produce optimal volume distributions with simple portal arrangements. Tolerance during treatment and, even more so, the lack of late subcutaneous fibrosis, is striking.

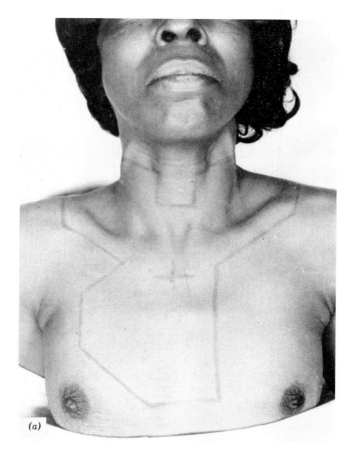

Figure 12. In many of the bronchogenic carcinomas, the lower neck should be irradiated. The treatment plan is best carried out by an anterior portal which includes the lower neck using 3 to 6 MeV. A given dose of 5000 rads is delivered. Optimally, to spare the spinal cord, the 25 MeV should be used for the posterior portal which covers only the primary and the mediastinum. (Reprinted, with permission, from Bloedorn, 1973)

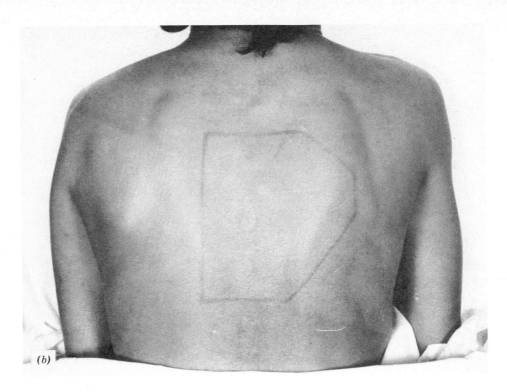

Fig. 12 (Continued)

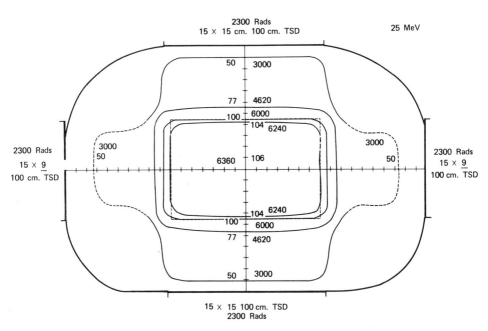

Figure 13. Whole-pelvis irradiation with four-field, "box" technique. The target volume is outlined by the dotted-line rectangle. This is a simple technique for patient set-up and daily reproducibility. The field sizes are 15 × 15 cm, anterior and posterior portals and 15 × 9 cm, lateral portals, 100-cm treatment distance. A uniform dose is achieved throughout the target volume with a favorable ratio of tumor dose to subcutaneous dose at entrance sites.

REFERENCES

Allt, W.E. 1969. Supervoltage radiation treatment in advanced cancer of the uterine cervix. A preliminary report. *J Can Med Assoc* 100:792–797.

Bloedorn, F.G. 1973. Thorax: Radiation reactions of normal tissues. p. 591–605. *In* Fletcher, G.H. (ed.) Textbook of Radiotherapy. 2nd edition. Lea & Febiger, Philadelphia.

Fletcher, G.H. 1956. A clinical program to evaluate the practical significance of higher energy levels than 1–3 MeV. *Amer J Roentgenol* 76:866–894.

Fletcher, G.H. 1973. Textbook of Radiotherapy. 2nd edition. Lea & Febiger, Philadelphia.

Levitt, S.H. 1973. Characteristics of Megavoltage External Beams. University of Minnesota Press, Minneapolis.

Shukovsky, L.J. & G.H. Fletcher. 1972. Retinal and optic nerve complications in a high dose irradiation technique of ethmoid sinus and nasal cavity. *Radiology* 104:629–634.

Tapley, N. duV. 1976. Clinical Applications of the Electron Beam. John Wiley & Sons, New York.

Biological Considerations in the Use of High-Energy Beams

Eric J. Hall, Ph.D.,
Professor of Radiology,
Radiological Research Laboratory,
College of Physicians and Surgeons,
Columbia University,
New York, New York

In recent years it has not been fashionable to think in terms of the biological advantages of high-energy x-rays and electrons. We have been obsessed with the potential benefits (real and imaginary) of more densely ionizing radiations, in the confident hope and pious expectation that these new modalities will solve all problems. The biologic considerations in the use of high-energy beams include, therefore, all of the basic problems of radiobiology that have occupied the attention of experimentalists for more than half a century. In addition, there are new phenomena relating directly to the increased potential of the present generation of large accelerators, in particular, their high dose rates.

THE 6 R'S OF RADIOBIOLOGY

In experimental systems, a number of radiobiological processes have been identified which alter the response of a tissue when a dose of x-rays is given in a series of fractions rather than as a single exposure. They are:

1. *Repair of sublethal radiation damage*, which takes place rapidly and is complete within an hour or so.

2. *Repair of potentially lethal damage*, the term applied to that component of the radiation damage that can be influenced by post-irradiation conditions. This is complete by 5 to 6 hours after an exposure.

3. *Reassortment* of cells, which takes place between dose fractions; the variation in radiosensitivity of cells with phase of the mitotic cycle results in selective killing and a measure of synchrony.

4. *Repopulation* of the tissue by the division of surviving cells, which may be delayed or speeded up by the influence of homeostatic mechanisms.

5. *Radioresistance due to hypoxia*, which dramatically modified the response to x-rays; it has been demonstrated to be a dominant factor in some animal tumors, but its importance in human tumors is unknown.

6. *Reoxygenation*, the process whereby a dose of radiation causes cells that are hypoxic to become oxygenated. The extent and rapidity of this process varies extremely between different animal tumors, and of course, its importance in the human is unknown.

Fractionation of dose will be beneficial only if one or more of the above factors operate differentially between normal and malignant tissues, leading to less damage to normal tissues than tumors. There is some evidence that the capacity for repair of sublethal damage may be smaller in tumors than in normal tissues; this would mean that fractionation would preferentially spare normal tissues. On the other hand, potentially lethal damage is repaired in tumors, but (presumably) not in normal tissues; this factor tends to adversely affect the therapeutic ratio. Reassortment of cells into various phases of the cycle, in principal, can take place in both tumors and normal tissues; the situation is too complex, and our knowledge too meager for any differential to be exploited. Repopulation, too, is a property common to normal and neoplastic tissues, the only difference being the influence of homeostatic controls. The remaining factors, radioresistance due to hypoxia and reoxygenation, apply only to tumor cells. Table 1 lists the factors which apply for x-ray treatments of tumors and normal tissues.

For neutrons, the situation is much less complex since repair of sublethal or potentially lethal damage does not occur, changes in sensitivity through the cell cycles are

TABLE 1. RADIOBIOLOGIC FACTORS IN THE TREATMENT OF TUMORS AND NORMAL TISSUE

Radiobiological Process	X-rays & Electrons		Neutrons	
	Tumor	Normal Tissue	Tumor	Normal Tissue
Repair, sublethal damage	yes	yes		
Repair, potentially lethal damage	yes			
Reassortment	yes	yes	yes[a]	yes[a]
Repopulation	yes	yes	yes	yes
Radioresistance due to hypoxia	yes			
Reoxygenation	yes			

[a] Reassortment takes place to a lesser extent for neutrons than for x-rays, but it is, by no means, neglible.

reduced, and hypoxic cells are less of a factor because of the lower oxygen enhancement ratio (OER). The principal remaining factor, where differences between normal and tumor cells must be exploited, is repopulation. This will depend critically on the particular combinations of tumor and dose-limiting normal tissue involved. Table 1 also lists the fewer variables in neutron treatments versus x-rays.

We can only conclude that x- or γ-rays represent a most complex form of radiation, for either biological experiment or clinical treatment. If there had been a choice in 1897, the radiobiologist would not have recommended using them! With the six R's of radiobiology available for manipulation, it should be possible theoretically to tailor an ideal treatment to any given circumstance—but in 1975 we are still in no position to do so.

In an elegant series of experiments at the Gray Laboratory, Fowler (1975) has demonstrated that it is possible to design a fractionated x-ray treatment regimen which is highly effective in eradicating a particular mouse tumor with minimum normal-tissue damage, provided the experimental parameters, such as cell cycle, rate of reoxygenation, proportion of hypoxic cells, etc., are all known. When this is done, changing from x-rays to neutrons, or adding a drug which is a hypoxic sensitizer, results in no further improvement; the tailored-for-the-tumor x-ray protocol is already optimal. It is unlikely that enough experimental data will ever be available for any human tumor to allow a special treatment plan to be evolved in a comparable way. The more practical lesson to be learned from these studies of Fowler and others is that the variable results obtained with x-rays can be made much more consistent and repeatable, either by changing to neutrons, or by using hypoxic sensitizers. Since this symposium is about x-rays and electrons, to suggest a change to neutrons would be out of place, and in any case, is inordinately expensive and possible at only a few treatment centers. The alternative suggestion is the further development, and introduction into clinical practice, of drugs which selectively sensitize hypoxic cells. This represents, in my view, one of the most significant advances in radiation therapy in this decade.

Clinical protocols have been developed on entirely empirical grounds, and it would be incorrect, as well as misleading, to pretend otherwise. However, there are certain general radiobiological factors that are clearly of importance.

THE SHAPE OF THE CELL SURVIVAL CURVE

X-rays are absorbed by interactions of the incident photons with planetary electrons of the absorbing material, giving rise to fast-moving secondary electrons, which, because of the ratio of their charge to their velocity, are relatively sparsely ionizing. For the most part, mammalian cells are inactivated by x-rays as a result of multiple events occurring in them. This accounts for the characteristic shape of the x-ray survival curve for mammalian cells, which commonly has a broad initial shoulder before approximating to an exponential function of dose. This implies that, at low doses, x-rays are relatively inefficient at killing cells, since damage must be accumulated before lethality occurs. The shape of the survival curve over the low-dose range attracted sporadic attention in the early 1960's, but it is only in the last few years that attention has focused on this initial region of the survival curve, culminating in the Gray Conference of September, 1974. At that conference a bewildering array of survival curves was presented and discussed, and their shape described by a number of mathematical expressions. The most widely used was the alpha-beta model:

$$S = e^{-(\alpha D + \beta D^2)} \quad (1)$$

This expression can be derived in a number of ways on theoretical grounds, but as far as we are concerned here, it will be used simply because it proves to be an excellent empirical fit to experimental data. (No theoretical basis is intended or implied!) This α component describes the initial slope of the survival curve, and reflects cell inactivation in single events, while the β term describes cell inactivation by the accumulation of multiple events, which is therefore a quadratic function of dose. It is at once apparent to the most casual observer, and confirmed by a careful survey of the literature, that α varies widely between different cell lines and also between different types of radiation. The value of β remains remarkably constant, not only for different cells, but even for a change of radiation type from x-rays to neutrons.

Figure 1 shows two theoretical survival curves which result in about the same surviving fraction for intermediate doses (in fact, they are matched to be identical at 600 rads), but which have markedly different values for α and β, the single- and multiple-event components of cell inactivation. Survival curve A has a shallow initial slope and bends over more rapidly at higher dose. Survival curve B has a marked initial slope, i.e., α is dominant.

Suppose two cell populations, characterized by survival curves with shapes A and B, are subjected to a multifraction regimen consisting of 30 doses of 200 rads. Their response, in terms of surviving fraction, is illustrated in Figure 2. After 30 fractions, population B is depleted to a much greater extent than population A. Many small fractions cause greater biological damage when α (the single-event component) is dominant. Alternatively, consider a regimen consisting of four dose fractions of 900 rads. Populations A and B would respond to this regimen, as illustrated in Figure 3. This fractionated treatment has a greater effect on population A than B. A few large fractions cause greater biological damage when β (the multiple-event component) is dominant.

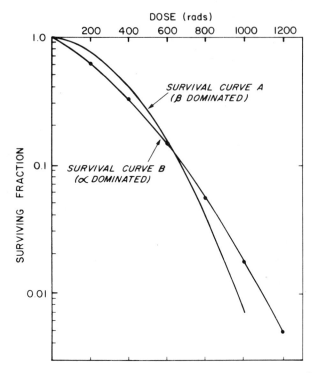

Figure 1. Two theoretical survival curves of the form $S=e^{-(\alpha D + \beta D^2)}$ with different combinations of α and β, but matched so that both have the same surviving fraction at 600 rads. Survival curve A has a shallow initial slope, i.e., this survival curve is dominated by the β or multi-event component. Survival curve B has a steep initial slope and in the low-dose region survival is dominated by the α component.

Depending on the survival-curve characteristics of the tumor and limiting normal tissues, many small fractions or a few large fractions could be advantageous and result in a favorable therapeutic ratio. There is infinite room for manoeuvre in this regard. To my knowledge, few, if any, radiobiological experiments have been performed with high-energy x-rays or electrons. Certainly none have concentrated on the shape of the survival curve over the low-dose range. It is not inconceivable that there could be some surprises in store. Much of the clinical experience with electrons, particularly from continental Europe, is hard to explain in terms of our present radiobiological notions, whereby everything from 50 kVp x-rays to 50 MeV electrons are lumped together, labelled as "sparsely ionizing," and assumed to have identical properties except for a small difference of RBE.

AN ANOMALOUS DOSE-RATE EFFECT

The dose-rate effect, commonly seen in a variety of biological systems, is readily understood in terms of the repair of sublethal damage occurring during the radiation exposure.

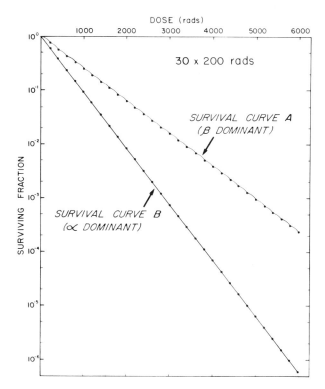

Figure 2. The response of populations characterized by survival curves of type A and B (illustrated in Figure 1) to 30 doses of 200 rads. For this fractionation regimen, population B is depleted to a much greater extent than population A, i.e., many small fractions cause greater biological damage when α, the single-event component, is dominant.

The biological effect of a given total dose of sparsely ionizing radiation is reduced when the dose-rate is lowered from hundreds of rads per minute, to tens of rads per hour.

The half time for the repair of sublethal damage (Elkind & Sutton, 1960) is approximately 1 hour, so that when the total duration of a radiation exposure is comparable with, or longer than, this half time, then sublethal damage will be repaired while the radiation is being absorbed. On this basis a dose-rate effect would not be expected when the dose rate is raised from 100 to 1000 rads/min, since in both cases the exposure time is short and the opportunity for the repair of sublethal damage during the exposure is minimal. This can, indeed, be demonstrated to be the case with cells cultured *in vitro*, and also with normal-tissue systems. However, it is not true of all biological systems, and, in particular, it is not true for the crypt cells of the mouse jejunum.

For this biological test system, a dose-rate effect has been clearly demonstrated by Hornsey and Alper (1966). Fast electrons delivered at a dose-rate of 6 kilorads/min were found to be more effective than x-rays at 500 rads/min in causing gastrointestinal death in mice (Fig. 4). The effect was clear-cut. In a later series of experiments, the somewhat crude endpoint of animal lethality was replaced by the more sophisticated endpoint of counting surviving crypts in the mouse jejunum. The conclusion was, however, the same; fast electrons at 8 kilorads/min were significantly more effective than x-rays delivered at 100 rads/min (Hornsey, 1970). These data are shown in Figure 5. This finding implies

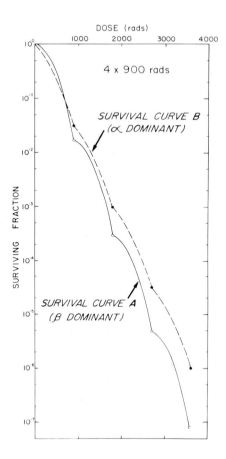

Figure 3. The response of two cell populations characterized by survival curves of shape A and B (illustrated in Figure 1) and their response to four fractions of 900 rads. This fractionated treatment has a greater effect on population A than B, i.e., a few large fractions cause greater biological damage when β, the multiple-event component, is dominant.

that there is a fast component of repair of sublethal damage evident in these cells that is not observed in other biological systems.

The jejunal crypt cells are unusual in several respects and certainly show a capacity to accumulate and repair sublethal damage which is unrivaled by any other biological system studied so far. Over the low dose rate range, the experiments of Withers (1972) and of Fu et al. (1975) indicate that the dose-rate effect is very much more dramatic for these cells than for any others that have been investigated. Figure 6 shows the variation in D_0 of the survival curve from very low dose rates by Fu et al. (1975) up to the extremely high dose rates used in Hornsey (1973); there is a dose-rate effect over this entire range. Comparable data for hamster cells cultured *in vitro* are shown in Figure 7. There is a marked dose-rate effect over the low dose-rate range, but above about 100 rads/min there is no further change of biological effect with dose-rate. In this case,

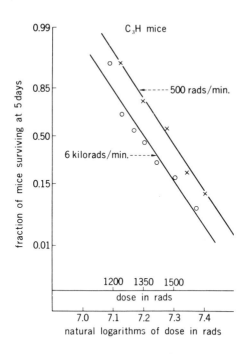

Figure 4. The percentage of surviving mice is plotted on an arithmetical probability scale against the natural logarithm of dose. The animals were of the C_3H strain (Taken from Hornsey, 1971).

the highest dose-rate used was approximately 10^{13} rads/min, obtained by delivering the radiation dose in an exposure time of a few nanoseconds.

With the same accelerator that Hornsey had used, Field and Bewley (1974) looked for a dose-rate effect between hundreds and thousands of rads/minute, by observing early and late reactions in the skin of rats. Their results are shown in Figure 8. When they increased the dose-rate from 200 rads/min to 6 kilorads/min, they found a reduction in the biological effect (in this case the average late skin reaction from 5-23 weeks). There was an even greater reduction when the dose-rate was further raised to 400 kilorads/min. This observation was readily explained in terms of local oxygen depletion occurring in the skin when the dose was accumulated at so rapid a rate. This explanation was substantiated by experiments in which the skin was irradiated under anoxic conditions, in which case the dose-rate dependence disappeared, as illustrated in Figure 9.

What are the lessons to be learned from these experiments in the context of recent developments with high-energy electron and x-ray accelerators?

1. In some well-oxygenated tissues, there may be a fast component of repair of sublethal damage leading to a dose-rate effect at high dose-rates, in the range readily obtainable with electrons and x-rays from the new generation of accelerators. This would lead to an *increased* biological effectiveness of the high-dose-rate radiation in some, but not all, tissues. In the human we do not know which tissues are involved, and anyone who

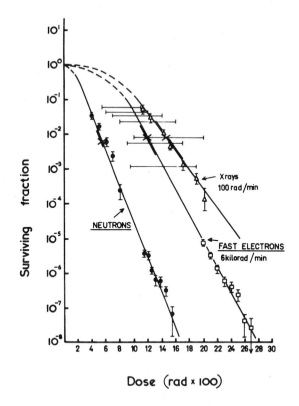

Figure 5. Epithelial cell survival curves estimated from crypt counts and clone counts in the irradiated mouse jejunum. Closed circles: irradiated with neutrons having a modal energy of 6 MeV at 50 to 100 rads per minute. Opened squares: irradiated with 7 MeV fast electrons at 6 kilorads/min. Opened triangles: irradiated with 7 MV x-rays at 100 rads/min. The crosses represent the LD_{50} (5) dose marked on the appropriate cell survival curves. (Taken from Hornsey, 1973.)

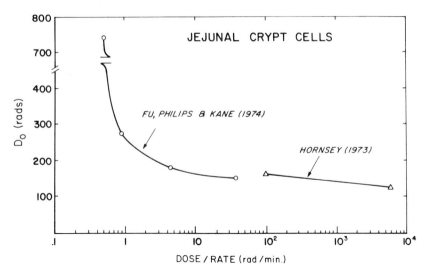

Figure 6. The variation of D_0 (the 37% dose slope) for the jejunal crypt cells in the mouse as a function of dose rate. The D_0 values over the low-dose rate range were taken from Fu et al. (1975); the high dose rate D_0 values were taken from Hornsey (1973). Over the entire dose rate range, the biological effect varies with dose rate.

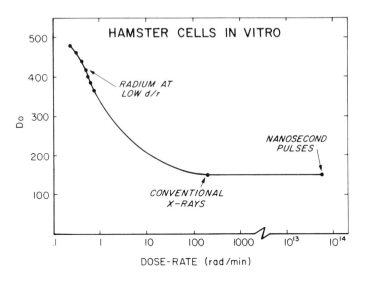

Figure 7. The variation of D_0 with dose rate for Chinese hamster cells cultured in vitro. Over the low-dose rate range the biological effect varies with the dose rate, but there is no difference in the biological effectiveness of 100 rads/min or 10^{13} rads/min. The highest dose rate was obtained by delivering the radiation in exposures of several nanoseconds using a Febritron at the Sloan-Kettering Institute in New York.

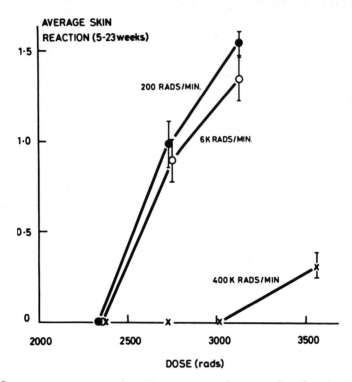

Figure 8. Dose response curves for skin reactions of rats irradiated with 200 rads/min, 6 kilorads/min, or 400 kilorads/min. (Taken from Field and Bewley, 1974.)

Biologic Considerations in the Use of High-Energy Beams

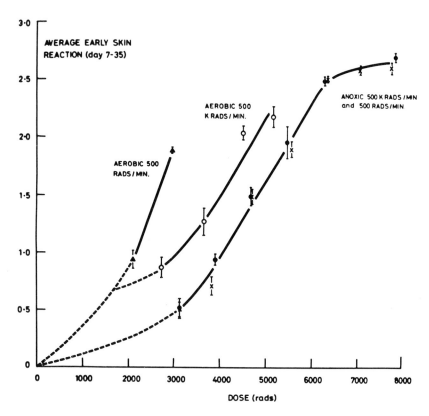

Figure 9. Dose response curves for the average early skin reaction of rat skin irradiated at either 500 rads per minute or 500 kilorads per minute, under aerobic or anoxic conditions. While there is a substantial dose-rate effect under aerated conditions, there is no dose rate effect under anoxic conditions. (Taken from Field and Bewley, 1974.)

pushes to very high dose-rates must be warned to be on the lookout for an anomalous dose-rate effect.

2. In some tissues, where a proportion of the cells are poorly oxygenated, such as might be the case in a poorly vascularized tumor, it is quite possible for radiation at a very high dose-rate to cause local oxygen depletion, which will result in a reduced biological effect.

In summary, it is not hard to visualize a situation in which increasing the dose-rate to a very high level may *increase* the normal tissue damage, and decrease the effect on the tumor. Since this is *not* the object of the exercise, increasing dose rates much above those conventionally used at the present time (several hundred rads/min) is contraindicated and should be viewed with extreme caution.

REFERENCES

Elkind, M.M. & H. Sutton. 1960. Radiation response of mammalian cells grown in culture. I. Repair of x-ray damage in surviving Chinese hamster cells. *Radiat Res* 15:556–593.

Field, S.B. & D.K. Bewley. 1974. Effects of dose/rate on the response of rat skin. *Int J Radiat Biol* 26:259–267.

Fowler, J.F. 1975. Dose fractionation schedules – biological aspects and applications to high LET radiotherapy. *J Canad Assoc Radiol* 26:40–43.

Fu, K., T.L. Phillips, L.J. Kane & V. Smith. 1975. Tumor and normal tissue response to irradiation in vivo: Variation with decreasing dose-rates. *Radiology* 114:709–716.

Hornsey, S. 1970. Differences in survival of jejunal crypt cells after radiation delivered at different dose-rates. *Brit J Radiol* 43:802–806.

Hornsey, S. 1971. Animal survival and cell survival seen in dose-rate effects on mouse intestinal damage. p. 207–212. *In* Sugahara, A. & O. Hug (eds.) Proceedings of the International Symposium on the Biological Aspects of Radiation Protection, Kyoto, 1969. Igaku Shoin Ltd., Tokyo.

Hornsey, S. 1973. The effectiveness of fast neutrons compared with low LET radiation on cell survival measured in the mouse jejunum. *Radiat Res* 55:58–68.

Hornsey, S. & T. Alper. 1966. Unexpected dose-rate effect in the killing of mice by radiation. *Nature* 210:212–213.

Withers, H.R. 1972. Cell renewal system concepts and the radiation response. p. 93–107. *In* Vaeth, J.M. (ed.) Frontiers of Radiation Therapy and Oncology. Vol. VI. Karger, Basel, and University Park Press, Baltimore.

The Value of and Need for High-Energy Electrons

John H. C. Ho, M.D.
Consultant in Charge,
Medical and Health Department,
Institute of Radiology and Oncology,
Queen Elizabeth Hospital,
Kowloon, Hong Kong

C. M. Lam, M. Sc.
Medical and Health Department,
Institute of Radiology and Oncology
Queen Elizabeth Hospital
Kowloon, Hong Kong

K. C. Lai, D.S.R.
Medical and Health Department
Institute of Radiology and Oncology
Queen Elizabeth Hospital
Kowloon, Hong Kong

INTRODUCTION

It has been more than a quarter of a century since high-energy photons from an American-manufactured betatron were first used as a radiation therapy modality in the management of cancer by Quastler et al. (1949). Soon afterward, betatrons were manufactured in Switzerland and Germany, and an intense enthusiasm was generated for the use of electron therapy in continental Europe by early favorable reports based on clinical impressions and also by an expectation of a possible specific selectivity of action of electrons on tumor tissue, as compared with other types of radiation. This expectation was based on the "two-component" theory, first propounded by H.H. Rossi (1961, 1964) of Columbia University, New York, and later developed by R. Wideröe of Switzerland (1965, 1966).

Although literature on the physical and biological aspects of electron therapy abound, reports on controlled clinical trials to check theoretical claims have been wanting. Clinical data available so far consist of those obtained from analyses of previously treated cases and clinical impressions derived from personal experience in the treatment of patients with the modality. At the Queen Elizabeth Hospital, Hong Kong, we have had a Brown Boveri 10- to 35-MeV betatron operating since 1964. Although it is capable of delivering both electron and photon beams, we have been using the machine solely for electron therapy. It is on the basis of experience gained in treating over 6000 patients with electrons alone, or in combination with high-energy photons from our linear accelerators, and a review of the experience of others that I shall attempt to assess in this chapter the value of, and the need for, high-energy electron therapy.

PHYSICAL CONSIDERATIONS

Depth-dose. Unlike photons, electrons with low energies have a sharp dose fall-off after a certain depth from the surface is reached, but with increasing energy the fall-off is more gradual (Fig. 1). There is only a slight build-up of dose beneath the surface due to scattering of electrons and photon contamination. It is the sharp dose fall-off characteristic that makes electrons of the lower range of energy eminently suitable for treatment through a single direct field of lesions at or close to the surface, extending ideally to a depth of up to 6 cm and no more than 8 cm.

Isodose curves. One disadvantage of an electron beam is that its edges are not sharply delineated, and there is a wide penumbra around the high-dose region (Fig. 2). Consequently, a generous margin of field coverage is required so that a target volume within the high-dose zone may be included. This makes electrons unsuitable for treatment through lateral fields of lesions close to the lens of the eye, e.g., frontal lobe tumors or those close to the brain stem, as in the case of nasopharyngeal carcinoma.

Tissue heterogeneity. When a non-unit density inclusion lies within the path of an electron beam, the uniformity of the scatter and absorption of the electrons is disturbed, and a non-uniform isodose distribution results. Figure 3 shows the disturbance caused by an air inclusion and Figure 4 that by carbon of density of 1.6 g/cc simulating bone. These inclusions give rise to "hot" and "cold" spots just distal to the edges of the inclusion. Heterogeneity of absorbing medium also causes disturbance of the isodose distribution of high-energy photons, but in a different way. For a long distance distal to the

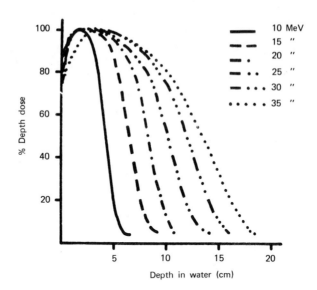

Figure 1. Central depth-dose curves in water of 14 cm X 14 cm beams of electrons of 10- to 35-MeV energies.

non-unit density inclusion, the dose is slightly increased in the case of air (Figs. 5 & 6) and slightly decreased in the case of carbon (Fig. 7).

Beam positioning. Another disadvantage of electron-beam therapy, using the conventional scattering foil device, as compared with high-energy photons, is that a long collimator is required to cut down photon contamination. This prevents the use of isocentric beam positioning, which is much more accurate than collimator positioning. In fact, for accurate electron-beam positioning, a well-fitting plastic shell with a platform or wax seating for the collimator is essential. Even with this help, the accuracy cannot reach that obtainable with isocentric beam positioning. Consequently, in situations where accuracy in beam positioning and shielding are required, as in the treatment of head and neck tumors, high-energy electrons are unsuitable.

With electron pencil-beam scanning, first described by Skaggs et al. (1958), the use of the conventional scattering foil and hence the collimator is dispensed with. The disadvantages of this technique are its high cost, due to its complexity, and the dependence of the dose distribution on the beam output during treatment, which should have minimal variation (Lanzl, 1968).

Intracavitary shielding. The ease with which electrons could be shielded by a thin layer of lead backed by a sheet of aluminum makes electron therapy of great value in the treatment of tumors of the buccal mucosa and lips. The tongue and gingiva can be readily protected by an intraoral shield. Similarly, the posterior wall of the vagina and, in the male, the posterior wall of the rectum can be shielded when the urinary bladder is treated by an anterior field.

Opposed fields. Figure 8-11 show the percentage dose variation from the midpoint to the periphery of a unit density absorber, when irradiated by a pair of opposed fields at different separation distances. It would appear from these figures that there is no place

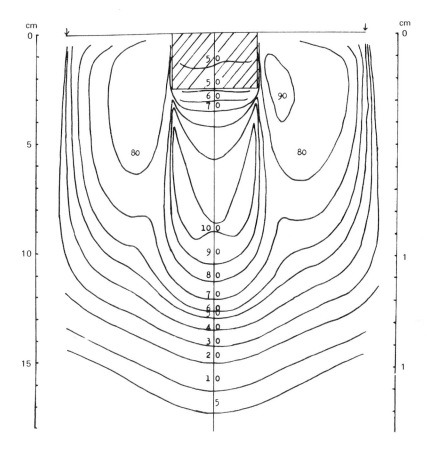

Figure 2. Isodose distributions of 8 cm × 6 cm beam of 35-MeV electrons in polystyrene (density, approximately 1 g/cc).

for electrons of the lower energy range for the treatment of a deep-seated lesion below the neck using opposed fields, because the field separation here is seldom less than 18 cm. For the treatment of intracranial lesions we have found 35 MeV, the maximum energy obtainable with our machine, to be the most used, either in pairs of the same energy, when the lesion is situated in the midline, or in combination with an opposed beam of a lower energy, when the lesion is eccentrically situated (Fig. 12). There is no real advantage of electron therapy over photons in the treatment of deep-seated lesions, whether opposed-field or pendulum irradiation, telecentric or otherwise, is used, unless the lesion is small, e.g., pineal or pituitary tumors. Even for these tumors a good dose distribution can also be obtained with high-energy photons by the use of a rotational field or three stationary fields. In most deep-seated lesions the target volume is, however, seldom small.

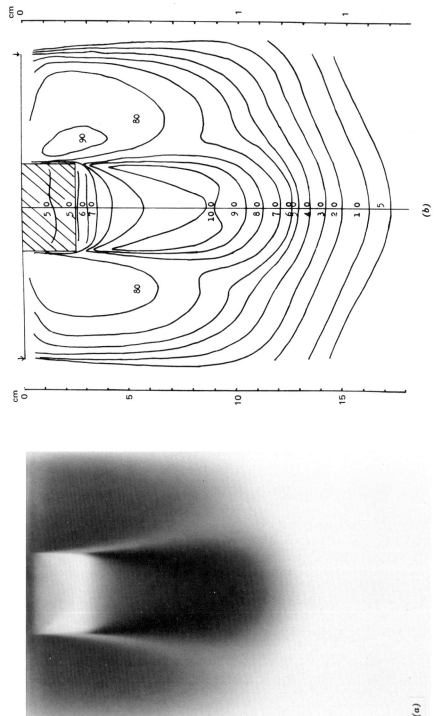

Figure 3. (a) Film sandwiched in perspex (1.18 g/cc), with an air cavity exposed to 35-MeV electrons. (b) Isodose distribution of Fig. 3a.

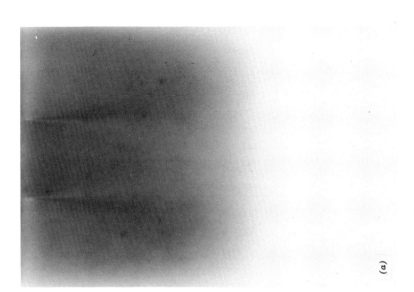
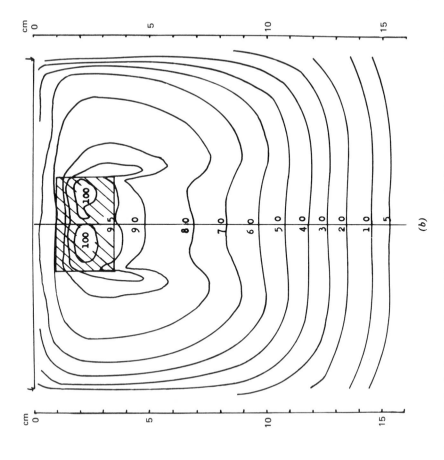

Figure 4. (a) Film sandwiched in perspex with carbon inclusion (1.6 g/cc) exposed to 35-MeV electrons. (b) Isodose distribution of Fig. 4a.

Figure 5. Film sandwiched in perspex, with an air cavity exposed to 4.5-MeV photons.

BIOLOGICAL CONSIDERATIONS

Skin tolerance. It is difficult to compare the data reported by various centers on skin tolerance in electron therapy, because it is difficult to measure the dose delivered to the superficial layer of the skin. Furthermore, the RBE correction factor for the r given by the machine is not uniform. We at the Queen Elizabeth Hospital use a correction factor of 0.83, for convenience, instead of 0.82, which was found by Huang et al. (1974) of our institute to be the RBE for 30-MeV electrons relative to 250-kV photons, using retardation of the growth of the tail vertebrae of new-born mice as the biological end-point. The use of 0.83 instead of 0.82 is because 0.83 is approximately equal to five-sixths, thus simplifying r-to-rad conversion. In practice, our clinical impressions tally with those reported by many others, principally Tapley and Fletcher (1965), in that the degree of skin reaction is between the reactions observed with kilovoltage and cesium-137 radi-

Figure 6. (a) A perspex head phantom with cut-outs simulating the nasal fossae, maxillary sinuses and the nasopharynx. (b) Film sandwiches in phantom, shown in Fig. 6a, exposed to 4.5-MeV photons.

ation. It is remarkable that skin reactions caused by electron irradiation tend to subside readily even after high doses, only to be followed several months later by subcutaneous fibrosis.

Soft-tissue fibrosis. We treated the necks of nasopharyngeal carcinoma patients in 1967 with a median lead shield of 2.5- to 3-cm wide through a posterior 14 cm X 14 cm field, using 35-MeV electrons to a total of 5 weekly doses of 700 r or 574 rads, reckoned at the 90% isodose level calculated to be equivalent to 1351 rets. This was followed by fibrosis, largely subcutaneous in location, in four of five patients followed up to 2 years. Four patients had four weekly doses of 574 rads, and none of them developed post-radiation fibrosis.

On the other hand, patients who were treated with 4.5-MeV photons from our linear accelerators developed fibrosis in only 4 out of 380 cases followed for the same period after receiving seven weekly doses of 560 rads at 90% isodose level through an anterior

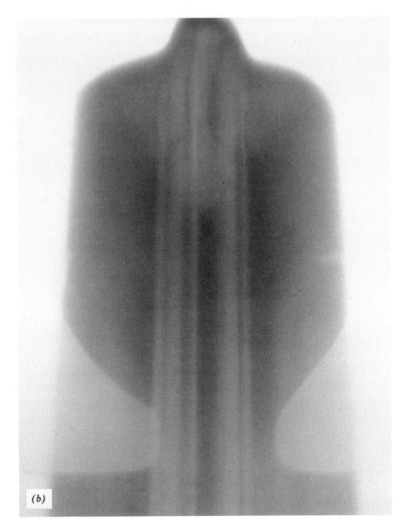

Fig. 6 (Continued)

field of 22 cm* X 12 cm. The nominal standard dose at 90% isodose level of this is 1629 rets.

Although the fields used and the dose distributions in the two groups of cases were different, it is felt that normal tissues are not spared any more in high-energy electron than in high-energy photon therapy. The fibrosis following the latter is always deeper and more diffuse, whereas in electron therapy it is more concentrated in the subcutaneous zone, especially where there is adipose tissue.

Electron electivity for tumor tissue. G.W. Barendsen (1964) published the survival curves of cells in culture given various doses of different types of low- and high-LET radiation. From his data Wideröe (1965, 1966) developed the "two-component" theory and speculated that electrons were superior to any other type of radiation with respect to an

---

*22 cm=width at the level of the supraclavicular fossae.

Figure 7. Film sandwiched in perspex, with carbon inclusion exposed to 4.5-MeV photons.

electivity of action on tumor tissue. It has now become clear from the body of evidence available that there is no apparent biological advantage of electrons over photons in this respect.

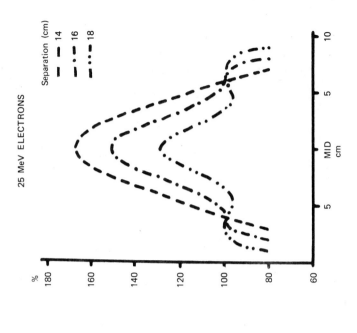

Figure 9. Central depth-dose curves of two 16 cm × 16 cm beams of 25-MeV electrons from opposite fields, with separation of 14, 16 and 18 cm.

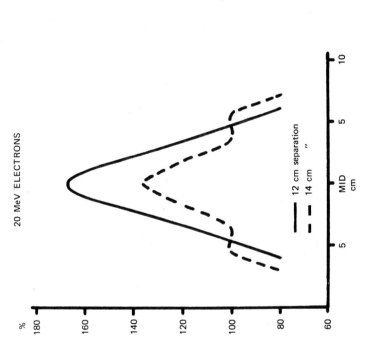

Figure 8. Central depth-dose curves of two 16 cm × 16 cm beams of 20-MeV electrons from opposite fields, with separations of 12 and 14 cm.

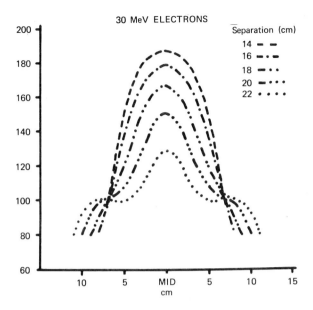

Figure 10. Central depth-dose curves of two 16 cm X 16 cm beams of 30-MeV electrons from opposite fields, with separations of 14, 16, 18, 20 and 22 cm.

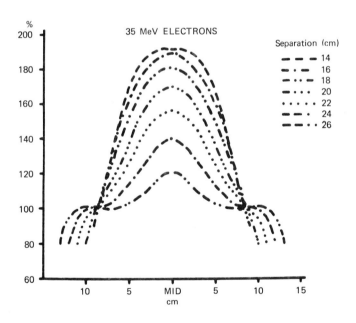

Figure 11. Central depth-dose curves of two 16 cm X 16 cm beams of 35-MeV electrons from opposite fields, with separations of 14, 16, 18, 20, 22, 24 and 26 cm.

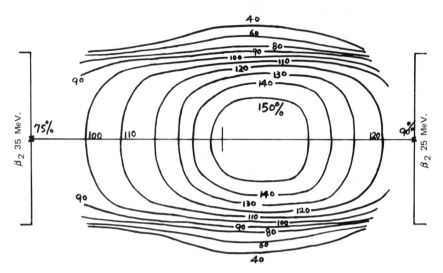

Figure 12. Isodose distribution of two opposed 8 cm × 8 cm beams of 35- and 25-MeV electrons, with 18-cm separation. Note eccentric isodose distribution.

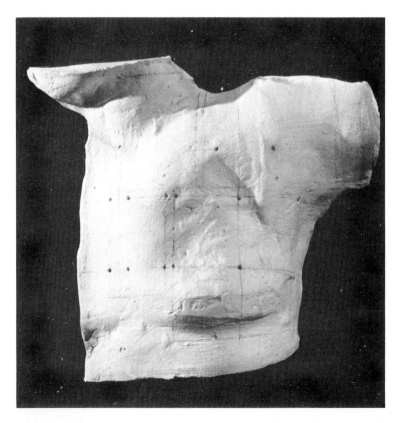

Figure 13. Plaster of Paris model of a patient who had a simple mastectomy. Note irregularity of surface contour.

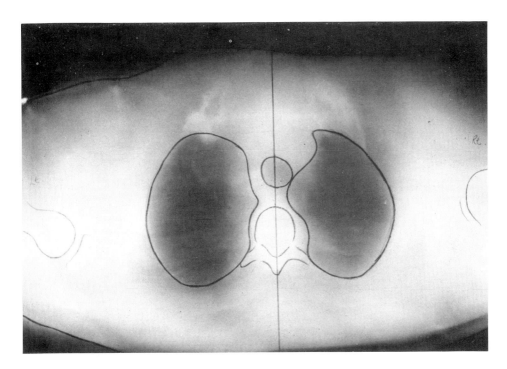

Figure 14. Transverse axial tomograms are required for determining chest-wall thicknesses at different levels. This one was at the level of the sternal notch.

THE NEED FOR HIGH-ENERGY ELECTRONS

From what has been said, the greatest need for high-energy electrons is for treatment of tumors at, or close to, the surface, with a distal target margin ideally not more than 6 cm, and at most not more than 8 cm, from the surface. The range of energy best adapted to the treatment of these tumors is from 6 to 25 MeV, and the technique used consists usually of a single, direct field.

Multiple adjacent fields, separated by a junctional gap of 0.5 to 1.0 cm, may sometimes be used if the area is large, as in the case of breast carcinoma after simple mastectomy (Figs. 13-16), but here electron therapy is normally used as just an alternative to other types of radiation therapy which are considered to be equally effective.

The situations where high-energy electron therapy is considered to be superior to other types from the point of view of physics, are listed below.

1. *Carcinoma of the middle ear.* Figure 17 shows the isodose distribution in the transverse plane at the level of the external auditory meatus for a 20-MeV electron beam delivered through a 7 cm X 7 cm field with a wax bolus compensator. The high-dose

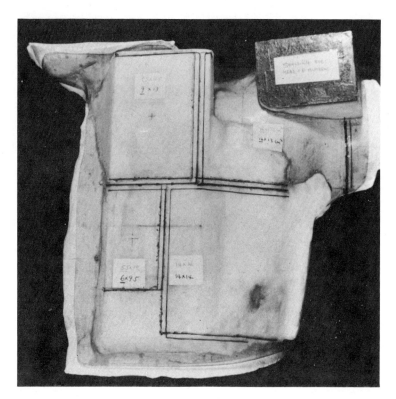

Figure 15. Cobex shell with wax bolus compensator, where the seatings for the four direct fields were marked. Note lead shield for the head of humerus. To minimize under- or overdosage at the junctional zones, the junction between two adjacent fields was moved 0.5 cm toward one field at one treatment and 0.5 cm toward the other in the opposite direction at the next.

region is encompassing the target area with only an insignificant dose contribution at the foramen magnum containing the brain stem.

2. *Carcinoma of thyroid.* The *lower part of the thyroid* is beyond the reach of lateral opposed fields because the shoulders and clavicle are in the way. An anterior field would bypass this problem. By having a wax compensator over the anterior median part of the neck, an anterior electron beam of 20-MeV energy can be made to deliver a high dose practically confined to the thyroid region and the anterior cervical nodes (Figs. 18 and 19).

3. *Invasive tumors of the face.* An illustration of the treatment technique for a tumor in such a location is shown in Figs. 20 and 21. The patient had a reticulum cell sarcoma of the subcutaneous tissue of the right cheek with extension to the side of the nose, inner canthus and upper lip on the affected side. An anterior 15-MeV electron beam was used. The left eye was completely shielded and the right eye partially, so that the medial wall and floor of the right orbit were included in the beam. So that the cornea and

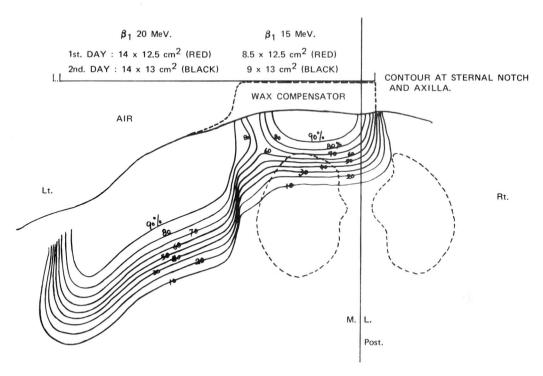

Figure 16. Isodose distribution at level of sternal notch and apex of axilla.

lens of the right eye would be spared, the eyes were directed to the right during the exposure.

4. *Tumors involving the anterior part of the nasal fossa or the anterior paranasal sinuses.* Because the upper part of the nasal fossae is situated between the orbits (Fig. 22), the eyes will be irradiated if opposed, lateral fields are employed. The problem in treatment planning is illustrated in the following case—a patient with carcinoma of the left ethmoidal sinus with extension to the nasal fossa and the frontal and sphenoid sinuses. One could only include all of them in the target volume in continuity, without irradiating both eyes, by employing an anterior field using electrons of the intermediate range, 15 MeV in this case, in addition to a pair of wedged, opposed, lateral, 4.5-MeV photon fields positioned behind the eyes, as shown in Figs. 23 and 24. If the maxillary sinus is to be included also, the ipsilateral margin of the anterior facial field should extend beyond the side of the face. This will cause a severe skin reaction at the pinna, unless it is shielded from the electrons eccentrically scattered from the face.

5. *Parotid tumors.* The advantage here is the sparing of the opposite parotid gland, thus avoiding the subsequent development of dryness of mouth.

6. *Fixed residual or recurrent peripheral lymph nodes.* Electrons are, perhaps, most frequently used for treating these lesions. For residual nodes, only a "boost" radiation dose is required. For recurrent nodes, great care must be taken to exclude the co-existence of a small or occult primary tumor, which may only be revealed by a biopsy of the region.

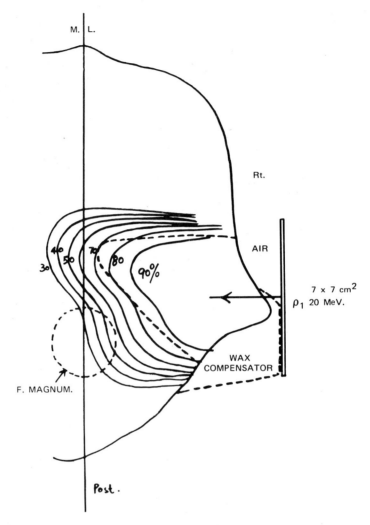

Figure 17. Case of carcinoma of middle ear. Isodose distribution in the transverse plane at level of external auditory meatus for a 7 X 7 cm beam of 20 MeV electrons with a wax bolus compensator behind pinna.

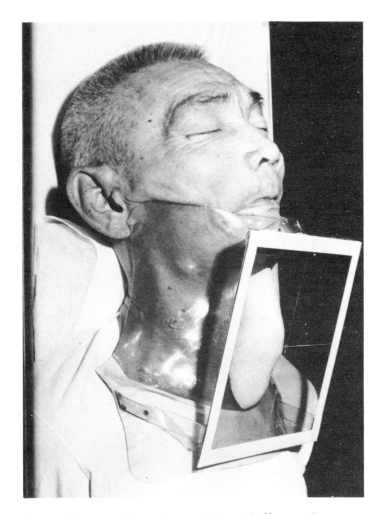

Figure 18. A case of inoperable carcinoma of thyroid. Note median wax compensator and collimator platform mounted on cobex shell worn over patient's neck.

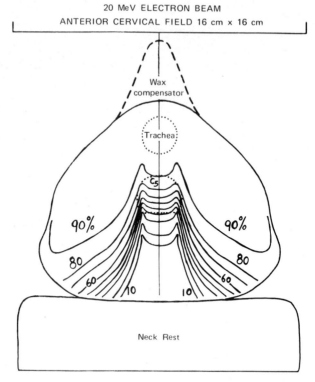

Figure 19. Isodose distribution in the transverse plane at the level of cricothyroid membrane.

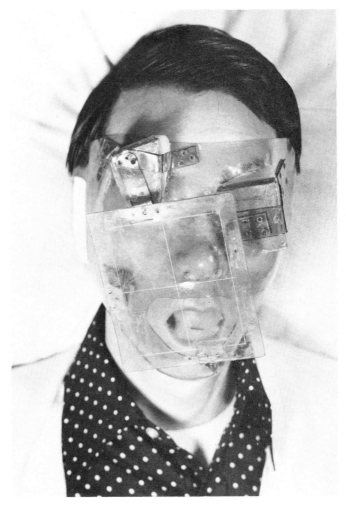

Figure 20. A case of reticulum cell sarcoma of right cheek (see text). Frontal view of patient wearing cobex shell with lead shields for the eyes and collimator platform.

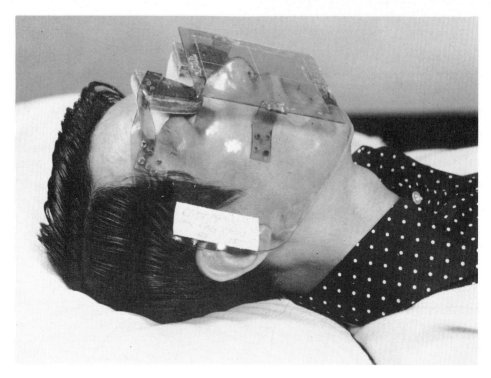

Figure 21. Side view of same patient as in Fig. 20.

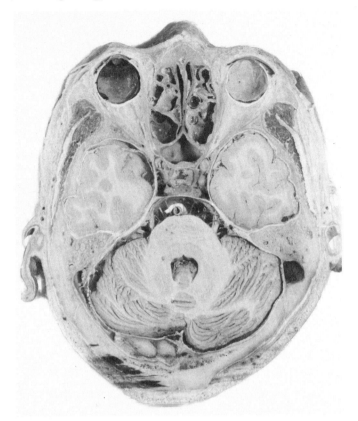

Figure 22. Transverse section of head at level of eyeballs to show relationship of the upper part of nasal fossae to the orbits.

Carcinoma *en cuirasse* may develop in the retreated nodal region if the persistent or recurrent primary is not treated as well. Interstitial therapy may be used as an alternative in the treatment of recurrent nodes, if they are superficial in location, but admission to a hospital, general anaesthesia and special expertise are required.

7. *Recurrent breast carcinomas on the chest wall and other superficial tumors or metastases.* These are normally treated by electrons lower than 10-MeV energy. Since the lowest energy of electrons given by our betatron is 10 MeV, we just add an absorber to reduce the energy to that, as required. This has the added effect of reducing the dose build-up beneath the surface.

8. *Anterior mediastinal tumors, e.g., thymomas.* These are eminently suitable for treatment by a direct, anterior, electron beam with a superior polystyrene wedge or wax bolus compensator. As some of these tumors may shrink during the course of treatment, frequent radiographic check on the width of the tumor is required. A narrower field may be required when there is tumor shrinkage. This would avoid undesirable radiation pneumonitis and fibrosis on the two sides of the tumor.

9. *Eccentric tumors of the oral cavity and oropharynx.* Some of these are better treated by interstitial therapy, e.g., carcinomas of the tongue that are small in size.

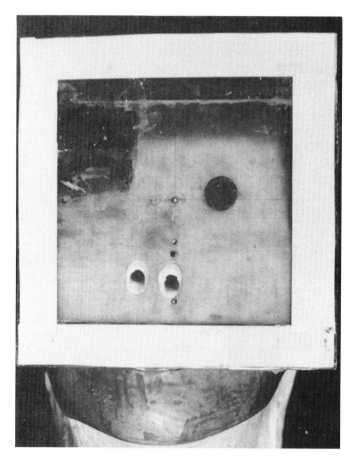

Figure 23. Frontal view of cobex shell with wax bolus seating and collimator platform for the anterior 15-MeV electron beam. Note nostril airways through the wax seating. A large lead shield for the uninvolved right eye and a small circular shield for the cornea and lens of the left eye were imbedded in the wax bolus.

The rest are suitable for electron-beam therapy, which allows a reduction of the radiation dose to parts distal to the lesion. Examples are carcinomas of the lip, buccal mucosa, lateral oropharyngeal wall and retromolar trigone. If more skin sparing is desired, electrons may be combined with high-energy photons of the 4- to 8-MeV energy range.

10. *Spinal metastases.* Here electron beam therapy is preferred because high-energy photons are too penetrating, and there is no skin-sparing with kilovoltage x-ray therapy.

11. *Urinary bladder tumor requiring external irradiation.* If the distal target margin is at more than 8 cm from the anterior skin surface, an anterior pelvic electron beam has to be supplemented by posterior, oblique, photon beams.

The situations where electron-beam therapy is considered a good alternative to high-energy photon therapy are deep-seated tumors with a target volume of not more than

Value of and Need for High-Energy Electrons

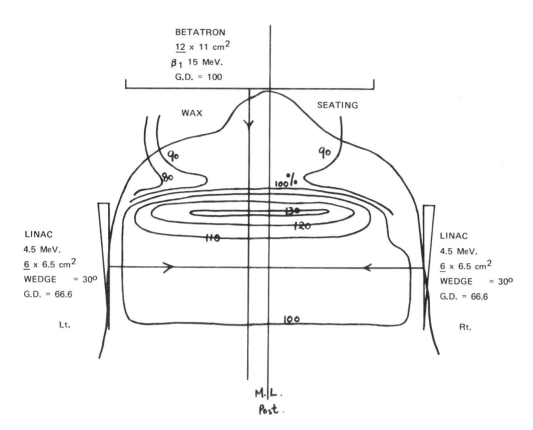

Figure 24. Isodose distribution at center of field in the transverse plane.

10 cm in its maximum diameter, preferably smaller, and treatable by opposed fields, separated by a distance of up to about 18 cm. Best examples are intracranial tumors, such as pinealomas, pituitary tumors, craniopharyngiomas, and medium-sized gliomas. Deep-seated tumors in the thorax and abdomen have a much wider distance between opposed surfaces at the level of the tumor, and usually a larger target volume as well.

CONCLUSION

There is a real need for high-energy electrons to cope with tumors in certain locations, where they may be used alone or in combination with high-energy photons. The latter are the bread and butter of a radiation therapy department. High-energy electrons may be considered in the same category as interstitial and intracavitary therapy, as playing a minor, but nevertheless an essential, role.

ACKNOWLEDGMENTS

Thanks are due to Mr. Richard Abesser for Figs. 1 and 8-11; Miss May Fong, Superintendent Radiographer (Therapy), and Mr. Andrew Tsui, Tutor Radiographer (Therapy), for extraction of treatment data; Mr. K. Leung for photography, and Mrs. P. Liu for secretarial assistance. Thanks are also due to Professor Peter Lisowski of the School of Anatomy of the University of Hong Kong for Fig. 22.

REFERENCES

Barendsen, G.W. 1964. Survival of human cells in tissue culture after irradiation with densely and sparsely ionizing radiation. p. 117. *In* Mammalian Cytogenetics and Related Problems in Radiobiology. Pergamon Press, Oxford.

Huang, D.P., G.F. Mauldon, C.M. Shen & J.H.C. Ho. 1974. Relative biological effectiveness (RBE) of 30 MeV electrons at varying depths in tissue. *Brit J Radiol* 47:795–799.

Quastler, H., G.D. Adams, G.M. Almy, S.M. Dancoff, A.O. Hanson, D.W. Kerst, H.W. Koch, L.H. Lanzl, J.S. Laughlin, D.E. Riesen, C.S. Robinson, V.T. Austin, T.G. Kerley, E.F. Lanzl, G.Y. McClure, E.A. Thompson & L.S. Skaggs. 1949. Techniques for application of the Betatron to medical therapy. *Amer J Roentgenol* 61:591–625.

Rossi, H.H. 1961. Measurement of absorbed dose as distributed in LET and other parameters. p. 1343–1348. *In* Trans. 9th International Congress of Radiology. Vol II. G. Thieme Verlag, Stuttgart.

Rossi, H.H. 1964. Correlation of radiation quality and biological effect. *Ann NY Acad Sci* 114:4–15.

Skaggs, L.S., L.H. Lanzl & R.T. Avery. 1958. A new approach to electron therapy (Proc. Second United Nations International Conference on Peaceful Uses of Atomic Energy, Geneva, Switzerland, 1958). *Isotopes in Med* 26:312, 904.

Tapley, N. & G.H. Fletcher. 1965. Skin reactions and tissue heterogeneity in electron beam therapy. Part I: Clinical experience. *Radiology* 84: 812–816.

Wideröe, R. 1965. The importance of newer radiobiological investigations for radiation therapy. p. 721–730. *In* Proc. XIth International Congress of Radiology, Rome, 1965. Excerpta Medica International Congress Ser. 105.

Wideröe, R. 1966. High energy electron therapy and the two-component theory of radiation. *Acta Radiol (Therapy)* 4:257–278.

Discussion

Dr. Robbins: Many years ago, a representative of the Brown-Boveri Company visited us and talked about the development of the betatron. He mentioned neutron contamination and the radiation hazard for personnel in betatron installations. He intimated that below about 30 MeV, there was not much problem; therefore he developed a 31-MeV betatron. In our own department a single survey reassured us that we were not at risk from neutron contamination at 25 MeV with the Allis-Chalmers betatron. Would you, Dr. Johns, care to discuss this problem?

Dr. Johns: What worries me is that some health organizations have laid down specifications for neutrons which almost make the production of one of these machines impossible. Every place I go I'm warned that neutron flux from these high-energy linear accelerators is a very serious problem. We have measured it and disagree.

Dr. Ho: Our physicists refuse to use brass collimator trimmers for our betatron applications, because of neutron production, even though our betatron, which has energies up to 35 MeV, is most often used in the 10- to 25-MeV range, where there is very little neutron production.

Dr. Almond: We have heard today that people have been treated for 25 years with betatrons which produced a measurable quanitity of neutrons. I have never heard any betatron therapist complain about late effects due to the neutrons or about what they do to the patients.

Dr. Johns: Is there any way this group can make recommendations so that we don't run into this problem? We all know that every patient, after betatron treatment, has radioactive carbon and nitrogen in them and will certainly activate a Geiger counter. But calculations I did many years ago convinced me that it was less than 1%.

Dr. Robbins: Dr. Fletcher, do you want to field your first question?

Dr. Fletcher: Dr. Fowler says that there is a difference in radiobiological equivalents (RBE) between cobalt-60 and 25-MeV radiation, so that when we say we give 7000 rads, it may not be quite 7000 rads. In Manchester when they had a 20-MeV betatron and the 4- or 3-MeV Metrovic linear accelerator, they did look into the matter. At that time they found no difference and reported this in the literature. Maybe there are some new data.

Dr. Leeper: I have some data that were gathered in the spring, at Jefferson. Our experiment used the Chinese hamster fibroblast system, and we plotted single cell survival as a function of radiation dose. Cells were irradiated with cobalt-60 gamma rays at 65 rads per minute, 45-MeV photons, and 45-MeV electrons from our Brown-Boveri betatron. Dose rate for the electrons and photons was about 110 rads per minute; for the cobalt-60, about 65 rads per minute. Statistically, there was no significant difference in the plotted curves for the three radiations. At 200 rads on the shoulder of the curve, where alpha becomes predominant, the cobalt-60 effect is, in fact, lower than that of the electrons or photons. But as this is not a significant difference, I conclude that there is no radiobiological difference between 45-MeV electrons, photons, and cobalt-60 gamma rays.

Dr. Hall: I would like to ask if this was all done in one experiment, with one common control? The difference between the two curves at the lowest dose, if it were all one pooled experiment, should be significant, but not if the experiments were separate.

Dr. Leeper: The electron experiment and the 45-MeV photon experiment each had its own set of controls. From the same lot of cells, on a given day, aliquots were irradiated with cobalt-60 or with either electrons or photons, but not both of the latter on the same day. The results were pooled so, in fact, the cobalt plot represents four experiments; the photon, two; and the electron, two.

Dr. Fowler: I agree that there is no significant difference between each of those points. However, if a curve were drawn through the points for cobalt and another through those for 45-MeV x-rays, I would expect statistics to show that the curves were different in the shoulder region, which is most important to fractionated radiotherapy. For the lower-energy radiation from cobalt-60, there is a steeper initial slope on the survival curve. It would certainly be right to do some further research along these lines.

Dr. Leeper: A common curve shows the similarities rather than the differences; until we actually extend our experiments and observations to the small dose up on the shoulder, I would not want to emphasize the differences.

Dr. Ho: My first question is, "In planning treatment and developing isodose plans, do you allow for air and bone surrounding the volume of interest?" When treating through lung, I have to take air into consideration, but I don't usually use electrons for treating through lung. In the case of bone, it depends on the type. For head and neck bones, which are quite compact, we have to use a bit higher energy in order to get through. But for the sternum, ribs, and other interior parts, we need not worry because they are rather thin. While the ascending ray of the mandible is certainly not compact bone, it is pretty tough, so we allow for that, too.

Dr. Hall: Dr. Fowler says, "You mentioned high dose rates but not much about lower dose rates, such as used in interstitial therapy or by Dr. Pierquin, that is, beam therapy at a low dose rate over a long time, eight hours or so a day. Do you think repeated small doses are biologically equivalent to continuous irradiation at a low dose rate? If so, high-energy machines might be used in short treatment times and simple setups two or three times each day."

Multiple repeated small doses should be equivalent to continuous irradiation, if we knew how many small doses were needed and how close together they should be. There is the difficulty. Unless someone uses multiple small doses instead of the Pierquin method, we will never get an answer.

Dr. Robbins: I think Dr. Coutard in 1930 practiced the technique of low-level radiation delivered at least twice a day, without concluding that it was superior.

Dr. Fletcher: This question was asked: "You indicated the clinical usefulness of mixing cobalt beams and a 25-MeV beam to attain acceptable dose distribution. Is there a place for beams at energies of about 10 MeV?" I do not see the point of spending money for equipment with a 10-MeV instead of a 4- to 6-MeV beam. It would be better to have a machine capable of selecting energies from about 4–25 MeV, such as the Sagittaire. In a department equipped with a 25-MeV linear accelerator and treating 30 or 40 patients a day, there must be at least two or three units. Treating a patient with a 4-MeV unit one day and with a 25-MeV unit the next, is cheaper in both equipment and maintenance.

That brings me to the question of field size in relation to energy. For 4 MeV, cobalt, or 25 MeV, the adjustment of field is the same, with about the same fuzziness at the edges. Exactly the same definition is used; the field is mapped and the beam obtained in the same way for each.

When you use electrons or photons, you must use a larger field (except for treating the thyroid, where the whole neck is involved and field size does not matter). There is very little constriction at 7 MeV; the curves are very flat. But the higher the energy, the more the constriction. When you get to 35 MeV, if you want to treat a 5-cm target volume, you must use close to a 10- instead of a 5-cm diameter. You lose a little in lateral tolerance, but you gain much because the opposite oral cavity or oropharynx gets no radiation.

Dr. Ho: The next question here is, "For a new installation, taking cost into account, what features would you recommend in order of need: high-energy beta beams, variable beta beams, or neutrons?" I would not want neutrons until I knew more about them. I have no neutron generator, so cannot speak with authority. The bread-and-butter of a radiotherapy department is a photon source. Having a photon source simplifies setup, which means you can treat more patients per machine. A photon source requires less computer calculation and so saves money. While the initial outlay may be higher, in the long run you could reduce your costs.

Next most desirable is a betatron, for electrons up to 25 MeV. Beyond that range there is some degree of neutron production. Also, most of the cases treated are eccentrically situated lesions, close to the surface. Having one machine to generate both photons and electrons is not wise, so you should get two photon sources and one betatron (for electrons). A betatron gives a fairly low output of gamma radiation, but a high output of beta. One little mistake might mean big trouble. I prefer to use each machine for one single purpose.

Dr. Hall: I have a question from Dr. Fowler. "Your warning about the possible adverse effects of very high dose rates is well taken, but from the evidence you presented, it seems safe to raise the dose rate by a reasonable factor above present levels. Do you think 1000 rads per minute would avoid the bad effects you mentioned? What about 2000, 4000, or 8000 rads per minute?"

It has been clearly demonstrated that there is a dose rate effect at the level of 8000 rads per minute. That is too high. For a number of reasons, 1000 rads per minute may be the practical limit. At more than that, horrendous effects may occur during an erroneous over-run.

Dr. Johns: Do you not have to specify the actual dose given in one pulse? A linear accelerator gives a nominal 1000 rads per minute. Is it actually that?

Dr. Hall: Here is another question which amplifies Dr. John's point. "The instantaneous dose rate during one pulse is very high for most accelerators, certainly in excess of a kilorad per second. What biological effects would you expect from this fact? Has any radiobiological work been done on this point?"

A few years ago ultra-high-dose-rate experiments were done, to look for a survival curve with the characteristic slope of hypoxia. With enormously high dose rates, the radiation used up local oxygen faster than it could be diffused in. Most of the experiments were performed with machines that gave the dose in a single pulse 1, 2 or 3 nanoseconds long. The group at Chicago worked with a high-dose-rate accelerator for which they estimated that during the time x-rays were effectively being produced (a very short part of the pulse), they were getting a dose rate as high as about 10^{12} or 10^{13} rads per second. They did cell culture experiments where surviving fraction plotted against dose gave a survival curve which broke to a slope characteristic of hypoxia. During the pulse they had a dose rate of about 10^{12} rads per minute in many of the pulses. There is momentary oxygen utilization in the pulse and, presumably, diffusion in between pulses. They did get the characteristic break, but they used quite a high dose, something like 500 or 600 rads as I recall. The most definitive work using nanosecond pulses came out of Sloan-Kettering, where Ed Epp, using bacterial systems, was able to show a clearly breaking curve. The break came at different dose levels, depending upon the oxygen concentration of the biological system. Dewey and Boag were the first people to propose the idea. Usually the break comes at doses above the daily dose encountered in radiotherapy.

Dr. Johns: There are several questions about the design of collimators for electrons. Until some basic work is done on electron-beam collimation, we will flounder around trying to build a good system. We need basic information, such as how electrons are scattered. Quite a few groups are now working on this problem.

Dr. Fletcher: Jack Fowler asks, "Can you comment on the potential usefulness [Well, I want more than potential; I want usefulness, period. You don't want to potentially cure the patient—you can do a little better than that!] of a machine to give variable energy up to only 15 MeV?"

Not useful. I see no point in going up only to 15 MeV. Along the same line, another question asks, "If you had the option to buy only one machine, would you buy an 18-MeV linac or a 45-MeV betatron?" The questioner reminds me that an 18-MeV linac has only 10-MV photons, which is correct. That is one of the reasons I object to the things. Further, the 45-MeV betatron has only 18-MeV x-rays, he says. That I don't know. However, the 25-MeV betatron gives a greater depth-dose than 4 or 15 MeV. If you have only one piece of equipment, you should use cobalt-60 (out of fashion, but still good) or a 4-MeV linear accelerator. That's your bread and butter. It meets most situations. If you have room for two (and, for that matter, enough patients), then unquestionably you should buy something which will give you a 25-MV photon beam. If you are rich enough and have sufficient patients, you can really begin to have optimal equipment. Your third machine should produce electrons. The Sagittaire can give you both photons and electrons; it is an exquisite gadget, but expensive.

A 10-MV photon beam, or linac, has disadvantages. The build-up is too deep. Many patients have superficial lesions, such as in the neck, supraclavicular, or groin nodes, where the tumor is mere millimeters beneath the skin. If you use only 10 MV for those cases, you will underdose the most superficial part of the cancer in the nodes. The 4 MV gives enough skin-sparing and will do for most cases, not optimally, granted, but better than any other. But you must treat tangentially, for example, when you use elective peripheral radiation for breast cancer or for the lower neck. Cobalt-60 gives the dose just under the skin in no time. The lymphatics which may be 2 or 3 mm under the skin, will be in the 90% dose zone. A 10-MV beam does not treat the lymphatics well at all, nor the low-neck tumors of the oropharynx or nasopharynx, nor the supraclavicular area in breast cancer. With a 10-MV beam, the build-up is too much, too deep.

With cobalt-60 or 4 MV, the nodes below the superficial nodes will get only 40 or 50% of the given dose. If you compensate by adding some build-up material, you ruin the skin-sparing. You cannot get the wax thin enough for both skin-sparing and sufficient dose build-up. Incidently, if you use 4 or 6 MV when you treat the breast with a tangential field, you have to use more bolus than with cobalt-60. You should experiment until you get a very deep, intense erythema at six weeks, with quite a bit of desquamation. That varies somewhat if you do lumpectomy; it is a little less with simple mastectomy for clinically advanced breast. (It is better to do a simple mastectomy than a radical, when there is lots of disease in the lymphatics of the dermis, right at the surface.) We always have to use some amount of bolus, even with cobalt-60, to create that deep erythema at six weeks, and for a desquamation. If you have 4, 6 or 10 MV, you have to use more and more bolus.

Most disease is really quite superficial. While 4-6 MV will not treat the patients optimally, if you have only one machine, that does better in a variety of situations. For your second machine, go to 25 MV; then, if you can afford it, get a newer machine to given both electrons and photons.

Dr. Johns: The problem of designating these machines should be clarified. When you talk about an 18-MV Varian, suddenly you realize it gives only 10-MV photons; that is a bit misleading. There should be two numbers given, such as, "Varian 18-10." That could mean it gives 18-MeV electrons and 10-MV photons. They should all be specified this way to avoid terrible confusion.

Dr. Fletcher: True. People who write me about equipment ask about that. You would be surprised at the number of people who don't understand. For them it is 18 MV, which is misleading. The manufacturers won't change the designation because it is such a good sales point.

Dr. Suntharalingam: There might be some misunderstanding, Dr. Fletcher. A 45-MV betatron gives 45-MV x-rays. I think the manufacturers have gone to the high energy, not to achieve any difference in depth-dose, but to get a higher intensity in the photon beam. If one looks at central-axis, depth-dose curves, there is very little to choose between 25, 35 and 45 MV. The reason the betatrons give x-ray beams, as well as electron beams, of up to 45 MV, is to get a flattened beam with a dose rate of close to 100 rads per minute at a 110-cm treatment distance.

Dr. Ho: I have a question: "Is there any place for rotational electron therapy in clinical applications? Is there any physical advantage or disadvantage in treating a curved surface with partial rotation with an electron beam? I understand that the French Sagittaire delivers electron beams by a scanning method."

For our machine, we use a long Perspex applicator which acts as a sort of collimator. This is applied directly to the part, just as when you use deep x-ray, orthovoltage, kilovoltage x-ray therapy. With that, you cannot have rotational therapy. The only time you can do it is with the scanning method, but I understand the dosimetry for the scanning method is extremely complicated and people are discarding it. For curved surfaces, such as the breast, I use adjacent fields having a gap of perhaps ½-1 cm, depending on how deep I want to go, and treating at a junctional area. One day the one field moves a little bit to one side, and the next day, the adjacent field moves to the other side, in order to iron out the isodose curves. I make a plaster model and then a cast, waxing up the uneven surfaces. The fields are moved so that the junctional area does not always have a gap. You must have transaxial tomography to help assess the depth of the lesion to be treated.

Dr. Fletcher: Peter Almond comments that in practice the amount of dosage is unimportant, since you can not get a patient out of the room, and the next one in and set up, in less than ten minutes. When you already give the treatment in as little as one minute, how can we increase the number of patients, even if we reduce the treatment time to zero? This is Peter Almond's comment, not mine.

Dr. Hall: Who suggested that increasing the dose rate would increase the number of patients?

Dr. Fletcher: It has been discussed so much, he wanted to throw a little practical note in that lofty discussion, that's all.

Dr. Johns: During Dr. Fletcher's talk, he implied that his linac gave the same depth dose as a betatron. Dr. Almond said they had different energies, and that the linac they were using in Texas had a thin transmission target, and the electrons were swept out of the beam. It is not surprising, therefore, that the depth dose was different.

Dr. Fletcher: I would like to paraphrase a verse of a famous French poet: "No matter the wine as long as it makes you drunk." No matter the energy as long as you get the same depth-dose. The depth-dose we get from the Sagittaire is the same as that from the Allis-Chalmers.

Dr. Hall: I have a question from Dr. Bannerjee of Pittsburgh, relating to an earlier question, namely, very high dose rates. "Has any experiment been done to prove conclusively that the high-dose-rate effect is due to depletion of oxygen available?"

Ed Epp has done experiments which proved beyond all reasonable doubt that the effect is due to oxygen depletion, at very high dose rates with nanosecond pulses. The break point of the curve varies predictably with oxygen concentration. At slightly less high dose rates, in the more practical range, the experiments of Field and Bewley showed a large dose-rate effect in rat skin under aerated conditions, which disappeared under hypoxic conditions. This clearly implicates oxygen in the process.

Dr. Fletcher: If I am correct, about 10 or 12 years ago, very high dose rates were used, about 10^{13} rads per minute, so as to bypass the oxygen effect. Hypoxic cells are already oxygen-depleted. If you use such a high dose rate with them, you will be more sparing of normal tissues. In other words, such a technique was a sort of tourniquet.

Dr. Podgorsak: Dr. Fletcher, your Sagittaire machine gives x-rays with no better depth-dose distribution than a linac. And you are using both a high-atomic-number target, and a high-atomic-number flattening filter. You don't advise us to use a 15-MV machine, and yet you are using the same machine.

Dr. Fletcher: I advise you to use a machine which gives you the depth-dose which my betatron gives me. I don't care about the energy.

Discussant: That betatron might not be all that you need, though. What kind of flattening filter are you using?

Dr. Fletcher: Sir, I want the depth-dose I have shown. I want to build-up at about 3-4 cm, and I want an 80% depth-dose at 10 cm, but the 10-MV linac machines do not give that depth-dose.

Discussant: But what about the 15-MeV linac?

Dr. Fletcher: No, that doesn't, either. As a matter of fact, one should avoid talking in terms of MeV's with any of the equipment. It would be best to define the equipment in terms of the depth build-up, and the percentage of maximum build-up dose at 10 cm.

Dr. Almond: The difference, I think, between Dr. John's machine and Dr. Fletcher's is basically this: Dr. Fletcher's machine has a transmission target, and lets through about 10 to 15 MeV of the electron-beam energy. The other is swept out by a magnet, so there is not a big target, but an individual target. This will affect the depth dose. Secondly, the energy on the target is around 26.5 MeV; we were comparing it to the 22-MeV betatron. Under those circumstances, the 10 cm X 10 cm fields for both machines match. There is also a difference in the energy; Dr. John's machine going from 25 to 60 MeV, while we were only going from 22 to 26.5 MeV. But the important thing is the fact that the target let through most of the electron-beam energy, which was swept out by the magnet. I agree with Dr. Fletcher; we can't just state the dose in terms of depth dose.

Dr. Fletcher: Each beam has advantages; the energy does not matter. You have a build-up at depth and a higher depth-dose, which is why the central dose is clearly higher with 25 MeV than it is with cobalt or 4 MeV. Therefore, it would be best to define equipment as giving a percentage of the maximum beam at 10 cm and ignore MeV.

Dr. Fowler: Dr. Fletcher, why do you set a 25-MV upper limit? One of the important aspects of the use of high-energy photons is the contribution of the photon beam to the biological dose in the region of the tumor. From a distribution of radiation, for pure dosimetry reasons, you showed that a 24- or 25-MeV betatron was superior to cobalt. If one calculates the biological dose to the regions away from the tumor, one comes to a better justification for high energy. We did several such calculations, and we have shown without any doubt that 25 MeV is not the limit. In fact, if you go above 25 MeV, even as high as 40 MeV, the ratio of the dose to the center of the tumor, compared to the dose outside the tumor, is much better. High doses probably will reduce bone-marrow damage, for example. A much more important factor than physical dosimetry is the depth of the patient. We have quite a few patients of about 20 cm on the anterior-posterior line; for these patients, the reduction of the dose to the off-center area is quite important.

Dr. Fletcher: Dr. Johns showed in his talk that over 25 MeV, the increase in depth-dose was negligible.

Dr. Johns: But you did get an increase; you cannot ignore it.

Dr. Fowler: You stated that most patients are treated with two techniques, while one technique, cross-fired, is the simplest. If you use cobalt, you cannot have one of a pair of fields in the vertical phase. It is better biologically to treat both of two opposite fields every day, as has been published in several papers, if you use the cobalt cross-fire technique. If you use a 25-MeV betatron, it is perfectly all right to treat one field day.

Dr. Fletcher: Yes, that is one of the advantages of a 25-MeV machine.

Discussant: I would like to comment on drawing higher x-ray energy with these machines. At least at 22 MeV, there is a tremendous amount of electron contamination due to pair formation, which reaches a depth of 10 cm. If you go higher in energy, this contamination becomes significant, and results in annihilation irradiation. You also get low-energy photon components. There must be a limit to the usefulness of higher-energy radiation in that light, even biologically. But electron contamination is very much a function of the machine's design.

Dr. Ho: Someone asks me, "Would you consider using 50-MeV electrons with parallel, opposed set-up for pelvic region?" I would not. The betatron producing the highest-energy electrons available to us goes up only to 42 MeV. Very little is gained between 25 MeV and 35 MeV, because you could readily substitute photons and be far more certain of the dosimetry. With electrons at that energy, we have a large penumbra, and the high-dose region is very small.

Dr. Fletcher: Here is another question: "You have shown examples of application of high-energy electrons and photons to head and neck, thoracic, and pelvic diseases. Have you found these modalities useful in abdominal malignancies—pancreatic cancer, for instance?"

If you could control cancer of the pancreas as you can cancer of the bladder, prostate, or cervix, of course you would do it. But no matter what we have given to pancreatic cancer (including neutrons, by the way), all the patients have died. This disease is too diffuse and metastasized, so that, in addition to killing cancer cells within the original volume, you chase the metastases.

Dr. Dobelbower: I have a comment, Dr. Fletcher. I am a radiation therapist at Jefferson Hospital. Pancreatic cancer is a disease in which the operative mortality exceeds the five-year survival rate. It often spreads, in the advanced state of the disease, to the liver, lungs or bone, by virtue of local extension, rather than metastasis. Radiation therapy has been largely palliative to date, but I think we have devised an effective treatment technique. We treated ten patients over the last nine months, and only one has died—from an infection, not from metastatic disease.

Dr. Fletcher: In 1954, when we got the betatron, we treated 10 or 15 cases of pancreatic cancer with parallel, opposed fields of 22 MeV. We gave them a fairly high dose, 7000 rads. Not getting very good results, we abandoned that tack, maybe too soon. Now we are trying again with neutrons. We have treated about five, and they have all died.

Dr. Dobelbower: With parallel, opposed fields, we find that other structures, including spinal cord and kidneys, receive a uniformly high dose through the posterior skin. We are taking the dose to 7000 rads, using two parallel, opposed 45-MV photon beams, and an anterior mixed beam (50% 45-MV photons and 50% 20-MeV electrons). We have varied the energy of the electrons, consistent with the depth of the lesion. This technique provides a sharp cut-off posteriorly, largely sparing the cord and kidneys—probably the most sensitive structures. In the ten patients treated thus, one has died from infection; the other nine are alive and well (except for one who has an intestinal ulcer).

Section II

Current Experience with High-Energy Beams

Current Treatment-Planning Practice with Photons

John S. Laughlin, Ph.D.,
Professor and Head,
Department of Medical Physics,
Memorial Sloan-Kettering Cancer Center,
New York, New York

In discussing current treatment-planning practice with photons, I shall deal primarily with a description of procedures developed and practiced at Memorial Sloan-Kettering Cancer Center, not that these are superior to those in other institutions but because they are the most familiar to me. Electron-beam treatment is also carried out regularly by our same treatment team, but these comments will be limited primarily to photons. This subject could also include implanted radionuclide treatment-planning, which is carried out with different physics personnel and procedures, but this subject has been recently reviewed in a similar conference (Laughlin et al., 1975) and is probably not directly pertinent to this meeting.

Although for decades diagnostic radiology had been able to use those radiation energies optimal for diagnostic purposes, it was not until after World War II that the high energies appropriate for radiation treatment were achieved. In addition to the primary impact of the associated advantages of high-energy x-rays and electrons, their development certainly facilitated the recognition of the entirely separate nature of the disciplines of diagnostic radiology and radiation oncology. With this increase in the capabilities of radiations available, as well as in the understanding of time-dose relationships, the importance of the treatment-planning process has been increasingly recognized.

The total treatment-planning process involves contributions by different personnel with different areas of expertise, operating as a team. The radiation oncologist, with the assistance of his physicist colleague and others, has to determine what is a desirable prescribed treatment and what is possible and to optimize the conditions for achieving the desired prescription.

TREATMENT PRECISION

A major concern is not only with optimization of the distribution of dose within the lesion relative to that in the healthy surrounding tissue, but with respect to the absolute magnitude of the dose delivered in the target volume. What absolute precision is desirable?

A commonly used guideline is that the tumor dose, however specified, should be within $\pm 5\%$ of the value prescribed. If consideration is given to all of the parameters, and emphasis is on the accumulated tumor dose actually delivered, this becomes a difficult level of precision to achieve in practice. This is particularly the case if the actual distribution of tissue density is not known or is not included in the dose-distribution calculation, if the precision in alignment in the repeated treatments is not rigorously controlled, if regression of tumor size is not considered in the plan, and if similar factors which make a difference from the total prescribed treatment are not taken into consideration.

The choice of a figure of $\pm 5\%$ as an objective is not entirely arbitrary. Observations of recurrence, on one hand, and complications, on the other, have indicated in certain instances studied that a variation in tumor dose of as much as 5% is deleterious. An analysis by Herring and Compton (1971) summarized the effect in patients of certain adverse complications, such as myelitis, resulting from radiation therapy. Similarly, analysis by Herring and Compton of data from Shukovsky (1970), who had reported on treatments of squamous cell carcinoma of the supraglottic larynx, showed that an increase of 10% in the nominal standard dose (rets) resulted in a drastic increase from less than 20% to 75% in the probability of tumor control, as indicated by lack of recurrence.

CATEGORIZATION OF THE TREATMENT-PLANNING PROCESS IN STEPS

The treatment-planning process described briefly here is a resume of the sequence of events classified in specific steps (see Fig. 1). Although this categorization is in 8 steps, it is recognized that, depending on considerations of weighting or semantics, a smaller or larger number of categories could have been employed to represent the process. Not included in this categorization are the regular machine parameter calibration and performance-checking procedures carried out on a regular basis.

Increasingly over the past two and half decades, the physical parameters have become those associated with high-energy x-rays and electrons from betatrons, linear accelerators and radionuclides, with energies ranging from a few million electron volts up to more than thirty million electron volts. It is interesting that the high-energy radiations from these sources are almost exclusively specified in modern therapy departments, which implies that the advantages of these radiations have been demonstrated to the satisfaction of the therapy community. Although there have been a few studies which have demonstrated

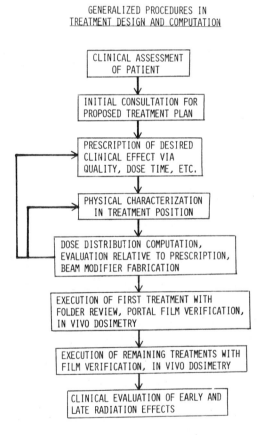

Figure 1. Schematic representation of generalized procedures in treatment design and computation.

the advantages of megavoltage radiation, even sufficiently rigorous for acceptance by biostatisticians, the universal acceptance of megavoltage radiation has been to a large extent based on recognition of the dose-distribution advantages and "clinical judgement," rather than on the results of specific clinical trials.

The frame of reference of the treatment-planning process has rested heavily, not only on what was clinically and biologically desirable in a given diagnostic situation, but also on what was possible within the limitation of the various parameters of the radiation apparatus available, within the limitation of the knowledge of the spatial and density distribution of different tissues, and within the limitations of which this information could be analyzed and integrated. These are briefly outlined below. Also mentioned because of its importance is our procedure for obtaining fundamental dose distributions under standardized conditions. In view of the volume of cases and the commitment of our staff to achieve essentially independent, individual planning in each case, it has been necessary to develop procedures which could and have been carried out on a large-scale basis. The physical aspects of this treatment-planning process are carried out by a staff of four junior physicists working under the immediate supervision of Dr. Simpson.

1. *Clinical assessment of the patient.* This includes the referral of the patient to the radiation oncologist and the initial examination and accumulation of diagnostic information.

2. *Initial consultation for proposed treatment plan.* The radiation oncologist, with the active participation of the clinical physicist, formulates a plan of attack which requires organization of the available clinical, radiobiological, and physical skills and radiation capabilities to achieve an optimum individualized course of treatment for the patient. Such questions as the following must be considered and answered: approximate patient dimensions at the tumor site; tumor volume; critical radiosensitive or dose-limiting tissues or organs adjacent to or in the treatment site; whether there will be or has been surgery at the treatment site, current or past chemotherapy, any previous radiation therapy, and if so, how much and where, so that its residual effect to both tumor, normal, and radiosensitive organs may be estimated; magnitude and uniformity of dose required in the tumor; what dose will be accepted by normal, radiosensitive or dose-limiting tissues or organs, etc.

As a result of these considerations, a treatment approach is developed which includes such factors as: beam quality; beam modifiers, such as wedges, blocks, compensators, or bolus; and positioning of the patient. Immobilization devices must be considered if there is any question about the reproducibility of the patient position. Adequate facilities for the fabrication of masks are important. If tumor-volume change is anticipated during the course of treatment, this should be considered at this point, since it will imply changes of the treatment plan during the course of treatment. The preferred time-dose fractionation scheme would also be determined at this time. Questions of *in vivo* dosimetry checks and documentation of alignment would also be specified.

3. *Prescription of desired clinical effect and associated radiation quality, dose, and time, etc.* As a result of the foregoing, the prescription is formulated by the oncologist, covering all of the parameters which need to be implemented.

4. *Physical characterization in proposed treatment position.* In this step measurements are carried out specifically in the proposed treatment position. These measurements yield the following: patient skin contour in one or more planes, usually transverse; determination of set-up marks relative to machine reference planes; tumor volume and location of

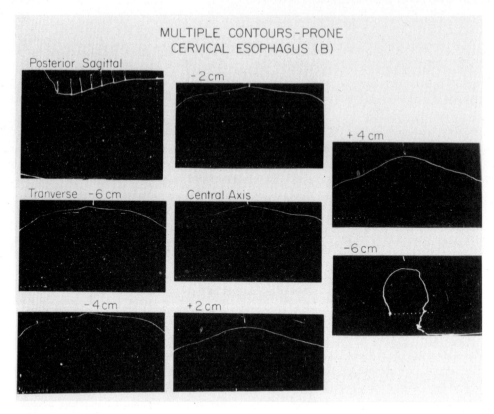

Figure 2. Multiple, parallel, transverse patient contours taken with ultrasound scanning arm at each of the levels marked on the sagittal scan. These data were analyzed to produce compensators for the prone treatment of a cervical esophagus case.

adjacent normal tissues which may be dose-limiting; location, extent, and nature of all tissues in the treatment region, particularly lung, bone and fat. The equipment employed at this stage includes contour devices, simulator, transverse axial tomography, and B-mode ultrasonography. We have also commenced use of the EMI computerized transaxial tomographic unit. Acquisition of a whole-body transaxial unit has been arranged and will be employed part-time for radiation treatment-planning.

As an illustration, in Figure 2 is shown the determination of patient contours obtained with B-mode ultrasonography. The upper left contour is posterior sagittal with 2-cm markers. At the level of each of these markers, contours in the transaxial plane have been obtained and are shown on the succeeding diagrams in Figure 2. These are obtained in preparation of a plan for an esophageal lesion. Present on each of these are 1-cm electronic marks used with an opaque projector to expand the polaroid copy to life size for computerized dosimetry and aluminum compensator design. In this instance the patient is prone and tilted up to make the esophagus level and in coincidence with the axis of rotation of the equipment.

Figure 3 shows another ultrasound view in the same case, indicating strong echoes obtained at the spinal cord. A typical example of breast-pleural interface is shown for a

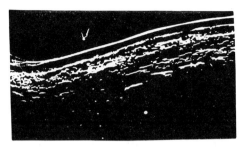

SPINAL CORD - PRONE CERVICAL ESOPHAGUS (B)

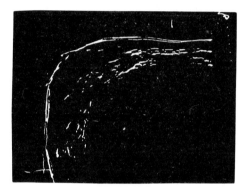

TRANSVERSE BREAST - PLEURAL INTERFACE

Figure 3. B-mode ultrasound scans typical for treatment-planning purposes. The upper scan is a midline sagittal posterior scan over the upper thoracic spinal cord (strong parallel echoes at depth). The lower scan shows a typical, right lateral transverse, thoracic scan for planning treatment of inoperable breast carcinoma, showing pleural interface.

separate case. The definitiveness of ultrasound as a tool for locating the extent of heterogeneities is, at best, presently marginal.

Figure 4 is the transaxial tomograph of this esophageal case, which is employed primarily to obtain the contours of the lungs. As has been previously noted, failure to correct for the lack of attenuation by the lung can cause the esophageal dose to be as much as 20% higher than prescribed.

A very effective method of obtaining patient skin contours is the use of a probe whose motion is recorded in two dimensions and is digitized for direct entry into a computer. Such a pantographic type of device has been in use in the radiation therapy department at the Gustav-Roussy Cancer Institute in France for some years. A convenient version, in which the locus of the probe is simply traced on paper, has been developed by D. Jones of the Northwest Medical Physics Center for distribution to their participating hospitals.

In cooperation with the Numonics Corporation, Memorial has developed a somewhat similar electromechanical, patient-contouring mechanism. This unit utilizes a Numonics graphics digitizer suspended over the patient with pantographic arms. The digital x and

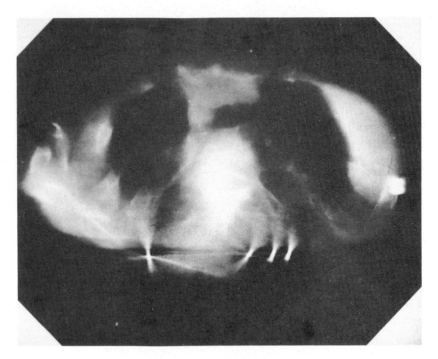

Figure 4. Transaxial x-ray tomograph of upper thoracic site for the cervical esophagus plan showing the spatial distribution of apices of lung.

y coordinates of the patient contour in multiple, parallel, transverse, or sagittal planes are accurate to within $\pm$ 0.5 mm over a total reading area of 50 cm X 40 cm. The system will not only be interfaced to display its results on an x-y system, but also with an analog plotter for true-scale contours and on-line entry into our computer-based treatment-planning system.

We anticipate that whole-body, computerized, transaxial tomography (CTT) will be extremely valuable to the treatment-planning process. In Figure 5 are two CTT views of a glioblastoma, which were employed in tumor localization and planning for this case. These views show a right parietal tumor extending into the occipital and posterior temple and posterior frontal regions. The ability of this method to localize seeds at the same time as the corresponding anatomical features will make this unit important for implant dosimetry treatment-planning, as well as for external radiation planning.

5. *Dose distribution, computation and evaluation relative to prescription, including beam modifier design and fabrication.* This step includes the computation of the treatment plan in as many planes as desired by the oncologist, with subsequent design of any wedges or differential absorbers needed. This step requires the availability of computer facilities and programs specific to the radiation apparatus being employed. It is necessary that the computational facilities be such as to permit quick generation for an immediate review of alternative-treatment-dose distribution in three dimensions. These dose distributions must accurately accomodate the effects of skin curvature, bolus, compensators,

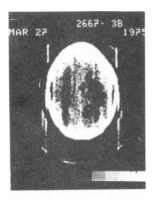

GLIOBLASTOMA-MULTIFORME

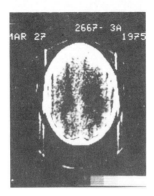

Figure 5. Computerized transaxial tomographic (CCT) scans generated on the EMI scanner for localization of a glioblastoma at two levels. Shows the large area of diminished density occupying the posterior right two-thirds of the right cerebral hemisphere.

wedges, finite source size, source distance, field size, tissue inhomogeneities, stationary and moving blocks, etc. If desired by the oncologist, the program should also make possible the display of the effects of time dose and fractionation to yield "isoret" distributions. Following selection of the optimal dose distribution for treatment, the associated beam modifiers are specified and constructed with the particular machine geometry in mind.

Figure 6 is the computed distribution of dose to be obtained with three fields, as indicated for the cervical esophagus case. The transaxial level is "central axis," as indicated in the contours of Figure 2. This diagram indicates the location, although not the design, of the compensators.

The dose throughout the target volume is uniform within 5% at all levels, and a lower dose to the cord has been specified between 30% and 50% of the tumor dose. Figure 7 is the "minus 4 cm" section. The apices of the lung are indicated, and correction has been made for their transmission on the assumption of an effective density of 1/3. This is an

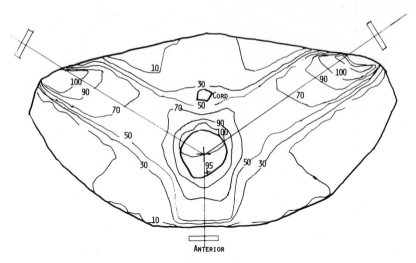

Figure 6. Three-field dose distribution in transverse section containing the central axis; corrected for curvature and Al compensator.

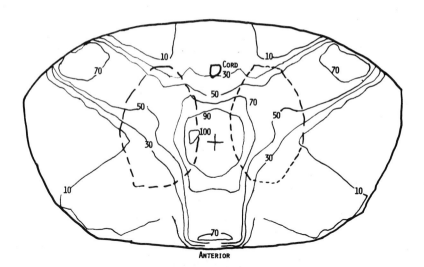

Figure 7. Three-field dose distribution in the lower transverse plane (central axis, minus 4 cm); corrected for curvature, lung heterogeneity.

estimated density. It can be determined by external scanning, as previously described, but this is not ordinarily done (Holodny et al., 1964). At this level the cord dose does not exceed 30% of the tumor dose, which is within 95% of the maximum. Figure 8 shows the "plus 6 cm" section with the cord dose approaching 50% of the target dose.

6. *Execution of initial treatments with folder review, portal film verification, in vivo dosimetry, etc.* Prior to execution of the first treatment, consideration must be

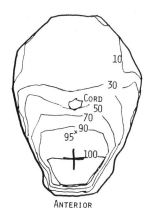

Figure 8. Three-field dose distribution in the upper transverse plane through the neck (central axis, plus 6 cm); corrected for curvature and Al compensator not shown in diagram.

given to whether or not the simulator should be used to verify the treatment ports and isocentric set-up markings. Therapy-machine, film-portal verification may be adequate for some treatments, such as parallel-opposed, single, or laterally opposed fields, but not for the case described here. Further, at this stage, a decision should be made as to whether *in vivo* dose verification may be desired at the target volume site and other adjacent critical sites, such as lens or gonads, and to check the accuracy of the fabrication of the blocking or compensators.

At this point it is insured that the documentation of the treatment-plan details, including dose distribution, are in the patient's record. In the cases where comprehensive treatment-planning has not been carried out, simpler treatment-planning procedures have, so that virtually all cases have been reviewed and dosimetry problems analyzed. In a busy department it is always conceivable that errors in implementation can occur, such as in timer calculations. An effective system has been the requirement that all patient folders be checked by the physics treatment-planning group before the first or second treatment. This is an independent check of any computations carried out, such as timer calculations, by the technologists, resident, etc. This feature of independent checking has proved by experience to be useful and important.

In vivo dosimetry is normally carried out by thermoluminescent rods or thin applicators, and on some occasions with ionization chambers. An instance is the determination of male gonadal dose consequent to treatment for Hodgkin's disease. Tables 1 and 2 show gonadal dose measurements obtained with different fields employed. In this particular case, approximately 4,000-rad total dose was delivered in regions of the thorax, spleen, para-aortic nodes, and inguinal nodes, with a total testicular dose of 500 rads. Such measurements are a part of the patient's record. It should be noted that Drs. E. Hahn and G. D'Angio (Feingold et al., 1975) are studying the radiobiological significance of these gonadal dose contributions in these patients and the effectiveness of surgical and physical methods in minimizing this dose.

7. *Execution of remaining treatments with film verification and in-vivo dosimetry.* This seventh step refers to the important area of reproducibility of repeated treatments.

TABLE 1. *IN-VIVO* MALE GONADAL DOSE MEASUREMENTS

Typical thermoluminescent dosimetry report of testicular dose measured during large field treatment for lymphoma emphasizing large contribution of dose by the pelvic ports.

	Mantle (III)	Para/Spleen (III)	Pelvis (III)	15MeVe$^-$ Boost	Mantle/Spleen (II)	Pelvis (II)
Date	5/10/74	5/24	6/10	6/12	7/31	8/22
Weekly prescription (rads/wk)	276 X 4	276 X 4	275 X 4	300 X 4	200 X 4	250 X 4
Total prescription	2210	2050	2200	1200	2000	1500
SAD/modality	173cm/6MV	173cm/6MV	173cm/6MV		173cm/6MV	173cm/6MV
Field size/depth (cm)	39 X 37 d9.5	21 X 17 d9	20 X 16 15 X 17 d9	7 X 6	44 X 42 d9	29 X 18 21 X 18 d8
Gonadal dose (% of tumor dose)	1.1%	1.3%	9.0%	0.8%	0.9%	12.7%
Gonadal dose (rad)	24	27	199	9	19	222

Total Gonadal Dose   500 rad

Current Treatment-Planning Practice with Photons

TABLE 2. *IN-VIVO* MALE GONADAL DOSE MEASUREMENTS

Testicular dose report for upper body treatment of lymphoma (mantle and paraaortic/spleen only).

	Phase I Mantle	Phase I Para/Spleen	Phase II Mantle	Phase II Para/Spleen
Date	11/7/74	11/21	12/5	12/16
Weekly prescription (rads/wk)	6 X 4	276 X 4	250 X 4	250 X 4
Total prescription (rads)	2208	2208	1500	1500
SAD/modality	173cm/6MV	173cm/6MV	173cm/6MV	173cm/6MV
Field size/depth (cm)	32 X 42 d11	21 X 20 d11	32 X 42 d10	22 X 20 d10
Gonadal dose (% of tumor dose)	1.3%	0.9%	0.5%	0.6%
Gonadal dose (rads)	29	20	7	9
		Total Gonadal Dose	65 rad	

Checks are made on the reproducibility of alignment, and remeasurements of patient dimensions and target volume can also be made to reassess the treatment plan.

8. *Clinical evaluation of early and late radiation effects.* This refers to follow-up by the radiation oncologist of his patient. It should be noted that when all physical problems have been thought of, solved, and properly documented in a patient's record, the physical variables become an effective part of the clinical study of results.

Figure 9 shows annual statistics (1974) for some of the physical contributions to patient treatment plans. Column A indicates computation of final dose distributions: the number of new cases comprehensively planned that year. Column B refers to the irregular field computations in connection with all blocked fields, the minority of which are Hodgkin's and non-Hodgkin's lymphoma treatments. Column C refers to computations carried out to determine dose at specific points where complete dose distributions were not desired.

Column D refers to numbers of cases in which *in vivo* dosimetry was requested and performed, and it should be noted that this varies from year to year, depending on the number of new radiation treatment procedures initiated by the oncologists. Column E refers to the number of cases for which special compensators and blocks were designed. This does not include the lung block designs carried out for the Hodgkin's disease cases. Column F refers to the number of patient folders reviewed. This is a procedure which is carried out not later than after the first treatment and applies to all external radiation treated cases.

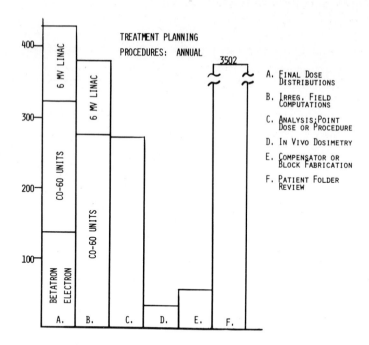

Figure 9. Annual statistical report of various treatment-planning procedures.

APPARATUS CHECK

The calibration of the output of the different radiation units, as well as a check on other parameters, is carried out on a regular schedule under the supervision of J.G. Holt. Calibration is made for a variety of field sizes and is done at the isocenter in a polystyrene phantom. These measurements are directly relatable to absorbed dose in water, and the exposure calibrations are directly traceable to the National Bureau of Standards. The mechanical accuracy of the isocentricity of the units is also checked, and the central-beam axis must be within a sphere of a few millimeters diameter throughout rotation. The correlation of radiation field and light indicators is determined, as well as uniformity of the field. Alignment lights, which are positioned on the walls of the treatment rooms for the isocentric set-ups, are also checked.

COMPUTERIZED TREATMENT-PLANNING DOSIMETRY

Important throughout the operation of the treatment-planning group has been the computerized systems and programs developed over the past two decades which make possible computation and analysis on a time-scale sufficiently short to be useful in the clinical planning process. We utilize a PC-12 with 16K core, disk drive, floating point processor, and other features, which is capable of carrying out computations for almost half of our external radiation cases. The PC-12 additionally serves as an intelligent terminal to a remote IBM 370 for those cases requiring a larger core.

The system is being continuously updated by Dr. Mohan and others of our staff. Other operations in the department that are important to treatment planning are also computerized. These include computer-aided measurement and analysis of dose distribution data, either with an isodensitometer or with an ionization chamber probe in a water phantom (isodosimeter). The isodosimeter has proved to be particularly effective in its design and operation. The system includes a water phantom, on one side of which is a thin (6/1000th of an inch) Mylar window to allow for measurement of the central-axis depth-dose (CADD) data at small depths. The CADD measurements are made with a thin-window, parallel-plate, ionization chamber of diameter 1.2 cm and a spacing 2.5 mm between the plates. The off-axis data are measured with a small cylindrical chamber of volume 0.05 cc. The ionization chamber is mounted on a drive mechanism and can be moved to any desired position by stepping motors with pulses received from the computer. The electrometer has two channels—one for the probe in the water phantom and the other for a reference chamber or a constant current source. Having moved the probe to a desired position, the computer sends the signal to the electrometer to trigger the integration of charge in the two channels, records the dose, and initiates the next motion and measurement cycle. Each cycle may take as little as a second. The measurements are reproducible to within one to two parts in a thousand. To obtain the maximum accuracy for a minimum number of measurements, the dose is sampled at shorter intervals where dose gradients are high, e.g., in the penumbra and electron build-up region.

ANCILLARY CONSIDERATIONS

In view of the importance of treatment planning, emphasis on dissemination of treatment-planning technology would appear to be important to effective radiation treatment. There is limited reason in spending large sums on megavoltage radiation equipment if the precision of which these units are potentially capable is not actually realized. Reviews in the field show that there is a wide range in the attention and expertise with which radiation treatment is carried out. Extension of apparatus and staff for modern treatment-planning procedures to all hospitals venturing radiation treatment would seem to be a productive endeavor. At this time, very appropriately, vast sums are being employed to examine the effect of high-LET radiation essentially for selected ambulatory cases. There is good reason to do this, but at the same time, for the same overall objective of the improvement of radiation therapy, it can be argued that major expenditure to extend treatment-planning technology would also be a very effective use of funds. One should take note that the establishment by the Cancer Control Program of six Regional Centers for Radiological Physics, and possibly more centers in the future, may provide a mechanism whereby such extension of technology could be facilitated. At present, however, the role of these centers is properly limited to consultation and monitoring assistance in those hospitals which have Cancer Control projects.

SUMMARY

Treatment-planning procedures, essentially as carried out at Memorial Sloan-Kettering Cancer Center, have been briefly reviewed. This necessarily incomplete resume of some of our treatment-planning procedures has been concerned primarily with physical aspects. It is suggested that since the government has given evidence of its desire to increase the effectiveness of radiation treatment, consideration of a deliberate extension of treatment-planning expertise more generally might make a sizable contribution in the interest of effective patient treatment, along with exploration of such innovations as high-LET treatment. In this connection it should also be noted that implanted radionuclides can often achieve a ratio of tumor dose—to—healthy tissue dose equivalent to that potentially achievable by high-LET radiation.

In conclusion, I wish to express my appreciation for the cooperation of Dr. G. D'Angio, Chairman of our Radiation Therapy Department and the other members of that department with whom we work, as well as to members of our own clinical physics group who have found real satisfaction in the development and employment of physics procedures in the interest of improved radiation treatment.

REFERENCES

Feingold, S.M., E.W. Hahn & L. Simpson. 1975. Aspermia and recovery following incidental gonadal irradiation in cancer patients. *Radiat Res* 62:610 (Abst).

Herring, D.F. & D.M.J. Compton. 1971. The degree of precision required in the radiation dose delivered in cancer radiotherapy. p. 51—58. *In* Glicksman, A.S., M. Cohen & J.R. Cunningham (eds.) Computers in Radiotherapy. Special Report No. 5,. British Institute of Radiology, London.

Holodny, E.I., G.D. Ragazzoni, E.L. Bronstein & J.S. Laughlin. 1964. Patient effective thickness contour measurement. *Radiology* 82:131—132.

Laughlin, J.S., L.L. Anderson & K. Pentlow. 1976. Aspects of implant dosimetry and problems. *Cancer*, in press.

Shukovsky, L.J. 1970. Dose, time, volume relationships in squamous cell carcinoma of the supraglottic larynx. *Amer J Roentgenol* 108:27—29.

Clinical Experience with High-Energy Photons

Thomas A. Watson, M.B.,
Director, Ontario Cancer Foundation,
London Clinic, Victoria Hospital,
London, Ontario, Canada;
Clinical Professor and Head,
Department of Therapeutic Radiology,
University of Western Ontario,
London, Ontario, Canada

Until 1925 no x-ray therapy equipment was capable of operating above an energy of 300 kV peak, but by 1933 a few very high-energy roentgen machines were available—for instance, the multisection tube at Pasadena, which operated at 1,000 kV peak, and a Coolidge cascade-type tube of 700 kV at the New York Memorial Hospital. These machines, however, were not robust enough for prolonged clinical trials. Apparatus of more than 2 MV did not become available for practical clinical work until 1948.

My own experience with very high energy, photon therapy started then when a 22 MeV Allis-Chalmers betatron (Kerst, 1943) was installed in the Physics Department at the University of Saskatchewan, Saskatoon (Harrington et al., 1949) as a cooperative venture with the Saskatoon Cancer Clinic, in which I had the privilege of working at that time. The primary function of this unit was physics research, but it was made available for biological and clinical work for a limited number of hours each day. Up until that time there had been very few radiobiological studies performed on photons of energy above 2 or 3 MV and no clinical work at all, except one terminal patient with a brain tumor, who had been treated by Quastler et al. (1949) early in 1948, the patient dying two weeks later.

In retrospect it may seem surprising, but at that time the sudden introduction of the clinical use of this radiation, since it was so much higher in energy than anything that had been used before, was approached with some trepidation and in the face of grave warnings as to the dire effects which might follow. Stone (1948) of San Francisco, was particularly perturbed, mentioning high-energy photons, specifically, because of the Californian experience of 1940 of the disasterous effect of neutron therapy. This preliminary use in patients was discontinued in 1943. However, there had been some work published by Quastler and Chase (1945) and Chase et al. (1947) suggesting that the RBE of these rays was in range of 0.65 to 0.8, depending on the biological material used, and this preliminary work was later reinforced by other studies of our own group. We could not see that any unusual hazards were involved. After all, biological effects followed ionization in the tissues, and the individual ions would be indistinguishable, however produced. The distribution of ions along the track of a photon was not very much different from that produced by the gamma rays of radium, which had been used successfully in the treatment of malignant disease for almost half a century.

Table 1 is reproduced from one of our early papers and follows concepts and calculations of Gray (1947), modified. The mean linear ion density per micron, as used at that time, has been now replaced, of course, by linear energy transfer (LET), but proportionately the implications are the same, and of course the LET of this type of irradiation is very different from that produced by neutrons, as pointed out at this time. It was also shown by Horsley et al. (1953) that the induced radioactivity in the tissue irradiated was of negligible significance.

TABLE 1. ION DENSITY FOR VARIOUS RADIATIONS

Apparatus	Energy (MeV)	Mean Linear Ion Density
X-ray generator	0.20	55.0
X-ray generator	3.00	8.3
^{60}Co unit	1.20	8.3
Betatron	22.00	6.7

Accordingly, in March 1949, we started treating our first patient without extensive and prolonged radiobiological formalities (Watson & Burkell, 1951). The selection of patients for several years followed a fairly consistent pattern, in that those suffering from advanced malignant disease of certain kinds, who, in light of previous experience, would have been extremely unlikely to respond to other forms of radiation treatment, were chosen. Only a few, in any case, could be irradiated because of the limited daily availability of the betatron and the geographic separation of the installation from our clinical facilities. This selection was further restricted by the fact that the beam was fixed in the horizontal position, thereby precluding the use of accurate beam-directed, small-field plans. In spite of these handicaps, however, we did succeed in treating a number of cases over the next decade or so and showed that, in fact, there were no unusual or unexpected hazards or complications (Watson et al., 1954). Several years later we reported on the clinical results of this early work and showed that a reasonable proportion of cases of Stage III and IV cancer of the cervix, for instance, could be cured by external irradiation alone (Watson & Burkell, 1959).

When we started treating patients on the betatron, we attempted to use techniques, in regard to fractionation, over-all time, volume irradiated, etc., in exactly the same way as we had been accustomed to treating similar cases with conventional 400 kV x-ray. By these means an attempt was made to change one parameter only—that is, the energy of the radiation used—in order to arrive at a clear idea as to the usefulness of this new modality. Because of the radiobiological experiments, it was realized that doses higher than those tolerable with conventional irradiation would be necessary, but the RBE varied according to biological material, and therefore an exact figure by which the dosage could be corrected was not available. Our original plan, however, was to treat our patients over a three-week period (shortly thereafter revised to five weeks), and tolerance levels for similar types of patients were soon established by clinical observation of, for instance, bowel, bladder, mucosal reactions, tumor response, etc.

A rather close parallel of these conditions exists today in respect to the imminent availability of pi-mesons for clinical use. The modality is new, biological data cannot be applied with confidence and exactitude to the human patient, the beam is fixed, and the output low. It will be interesting to see whether less empirical approaches are made to rather similar problems more than a quarter of a century later.

The introduction of cobalt-60 units in 1951 overshadowed developments in very high-energy photon work, and, because of their simplicity, reliability, reasonable output, flexibility, and cost, they were an almost instant success. Meanwhile, because of the technical difficulties in producing practical higher-energy apparatus, development of betatrons, synchrotrons, and linear accelerators, usually with electron capability, proceeded very slowly. Now, however, fairly satisfactory apparatus, albeit expensive, of energies up to 45 MeV, is available and in routine clinical use. Most of these machines, as an important dividend, have electron capabilities. The so-called "conventional deep x-ray therapy" has suffered an eclipse, and, in fact, in our own department, the last such machine was scrapped several years ago.

From the practical treatment point of view, the advantages of high-energy photon therapy have been so often documented that no elaboration is necessary. They include the production of very high percentage depth-doses in tissue, marked skin-sparing effect, a more equal energy absorption in different body tissues, lower integral doses for irradiation of specified deep volumes of tissue—all when compared with conventional deep x-ray therapy (Watson et al., 1954). When, however, these physical parameters for low-megavoltage photons are compared with those pertaining to high-megavoltage photons, modifications and qualifications to these statements have to be made. Thus, although the percentage depth-dose at, say, 10 cm, is still considerably superior for the very high-energy photons, the advantage is not nearly so marked as when low-megavoltage photons are compared with kilovoltage photons. Differential absorption in bone, compared to muscle, actually increases slowly at energies over 10 MV peak (Johns, et al., 1951; Watson et al., 1954) (Table 2). Although the lower megavoltage produced a marked

TABLE 2. ENERGY ABSORPTION PER ROENTGEN IN TISSUES

	Energy Absorption (ergs per gm)			Energy Absorption (ergs per cc)		
	Fat	Muscle	Bone	Fat	Muscle	Bone
200 kV	70	91	300	64	91	555
Cobalt-60	98	95	84	89	95	155
Betatron (25 MeV)	91	92	91	83	92	168
Density of tissues	0.91	1.00	1.85			

skin-sparing effect, there is not the same sparing of normal tissues at a somewhat deeper level. It would seem that any clinical advantage in the use of very high-energy photon beams arises from their peculiar distribution of dose in the tissues and not from any intrinsic biological advantage. Although there seems to be a slight difference in RBE, there is no reason to expect any difference in OER.

The question arises as to what the maximum deep energy of megavoltage apparatus should be in clinical use. At the present time, we have betatrons of 45 MeV, but nothing, as far as I know, of a higher energy is used for the production of photons for clinical use. For practical purposes, if we assume a maximum tumor depth of 10 cm and a total body thickness of 20 cm, we see from Figure 1 that the 100% dosage level will occur at about 10 cm when the MeV is 100 (J.C.M. MacDonald, Personal communication). At the same time, while the tissues between the surface and the 10-cm depth receive less than 100%, those between 10 and 20 cm receive close to 100%. If we assume the most adverse situation in which the tumor center is 10 cm in depth and 8 cm in diameter, including a 1-cm margin, it will be appreciated that there is probably a disadvantage in using a beam which produces 100% depth-dose at more than 6-cm deep. Tumors which are more shallow will be more suitably dealt with by lower energies. Thus, it would seem, for photon therapy, that megavoltage therapy should be restricted to a peak energy of less than about 40 MeV (Fig. 2).

There is a widespread opinion that the sharpness of the edge of the beam improves with higher energies. Thus, with cobalt-60 irradiation, there is a considerable geometric penumbra produced, depending to some extent on the size of the source, the skin-tumor distance, the source-tumor distance, and the collimation, including the distance from the last limiting diaphragm. If, however, the edges of similar beams are compared for cobalt-60 radiation and 34-MV photons (Fig. 3; D. Dawson, Personal Communication)

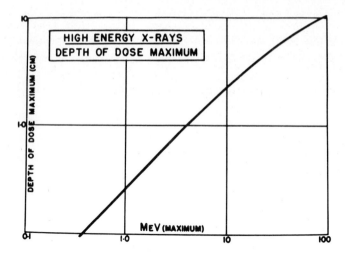

Figure 1. *High-energy x-rays—depth of dose maximum.*

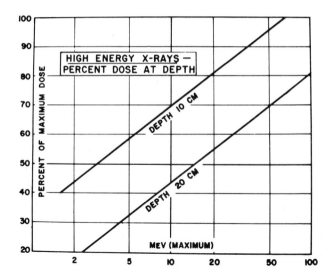

Figure 2. *High-energy x-rays—percent dose at depth.*

it will be seen that such opinions are largely illusory, and that while the diameter of the source of cobalt irradiation necessarily produces a larger geometric penumbra than that displayed by reason of smaller focal spots by very high-energy, accelerator-produced photons, the lateral transport of the energetic secondary electrons produced in the latter instance accentuates the penumbra effect. The result is that, at practical working distances and with presently available equipment, there is virtually no advantage, in this respect, in the use of the higher photons, since the penumbra is virtually independent of depth, field-size and field-elongation.

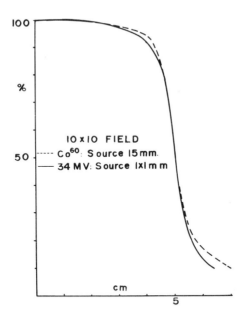

Figure 3. Penumbras in a 10 cm X 10 cm field irradiated by ^{60}Co radiation and 34-MV photons.

The terms "high-energy" or "megavoltage," as descriptive of the quality of photons being used, are too broad to be informative, since they can refer to energies of just over 1, up to 45 MeV. Furthermore, in conformity with x-ray therapy, the type of beam is described by peak energies, except in the case of cobalt-60 units. This hardly describes the spectral distribution in the beam. Some linear accelerators, for instance, with high peak energies, produce an effective beam comparable with betatrons of very much lower nominal peak energies (Podgorsak et al., 1975). A more scientific and objective descriptive convention should be used, as was the case in conventional "deep" x-ray therapy, when the half-value layer, in copper, was stated almost invariably. It is suggested that a parameter such as the depth at which the dose becomes a maximum (J.C.M. MacDonald, Personal Communication) might meet this need, or, alternatively, the percentage depth-dose at 10 cm, for a 10 cm X 10 cm field at that depth, (H.E. Johns, Personal Communication) a standard tumor-source distance (e.g., 100 cm) being used. The latter is probably easier to define precisely. Thus, the quality of the beam of a 4-MeV linear accelerator might be described as "D_{10}-63%" and that for a 35-MeV betatron as "D_{10}-88%.

Although the clinical use of these very high-energy photons is widespread, there is really very little objective, practical, published literature documenting improved effectiveness in specific situations. Allt (1969), however, has conducted a controlled clinical trial in which moderately advanced cases of cancer of the cervix were treated primarily by either cobalt-60 irradiation or 22 MeV x-rays. Although the techniques and dosages used in each of the two groups were not strictly comparable, a very marked superiority in the five-year survival rate was demonstrated in the patients receiving the higher-energy modality. It would seem to follow that, at least in the same location, other tumors, e.g., carcinoma of the bladder, rectum, and perhaps ovary, would be equally suitable for this kind of treatment. It should be pointed out that there are some disadvantages and difficulties in this kind of treatment. All beams have to be flattened by appropriate filters, and therefore the output is necessarily greatly reduced. Unless the original output is enormous, the flattening of the field usually precludes the production of fields larger than 20 cm × 20 cm. Furthermore, installation is expensive, and the machines necessarily large and somewhat clumsy.

In our present department we have had available, for almost 10 years, a 35 MeV Brown-Boveri betatron. Although electron beams of energy of 10 to 35 MeV are obtainable, the machine actually operates at 34 MeV only, in the production of photons. The apparatus is well-engineered and relatively easy to position, and 190° rotation can be obtained, with accuracy, by integration of the movements of the treatment couch and treatment head. Low-megavoltage apparatus, in the form of cobalt units, are also in use in our department.

Under these circumstances it may be of interest to survey the type of patient selection which has evolved in our own situation; Table 3 lists, by percentage, the different diagnoses involved. It will be noted that the majority of patients are suffering from some type of pelvic disease, in all of whom it is considered necessary to irradiate the whole of the pelvis, because of the stage and paths of actual or possible spread. Thus, patients with cancer of the cervix, Stages III or IV, or, in certain circumstances, Stage IIB, are irradiated with 34-MeV photons using a three-field plan. At this energy a four-field "box"

TABLE 3. PATIENTS TREATED BY 34 MeV PHOTONS BETWEEN 1966 AND 1974.

Site	Number	% of Total
Bladder	234	23.0
Cervix	198	19.5
Other pelvic tumors	167	16.5
Esophagus	115	11.3
Bronchus	196	19.0
Miscellaneous	107	10.5

is not necessary, as is seen from the distribution in Figures 4 and 5, in which the distribution of dose in the pelvis from cobalt-60 photons is compared with that obtained with both 6- and 34-MeV photons. This arrangement also spares the rectum to a fair degree. Usually a maximum tumor dose of 6000 rads, and minimum dose of 5400 rads, is delivered to the whole pelvis in 24 treatments over a 6-week period (four treatments per week). This is either followed, or preceded, by a local radium insertion, using the Manchester technique, to the uterus and vagina, to a dose of 1300–2600 rads at point A.

All of these patients develop some degree of diarrhea towards the end of the treatment course, and in some cases there is a cystitis. The commonest late complication is a constrictive sigmoiditis which has been reported as occurring in 10% of cases (Fletcher, 1964).

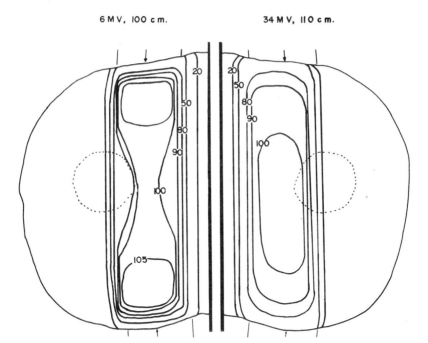

Figure 4. Comparison of radiation by 6-MV photons and ^{60}Co radiation, three-field plan.

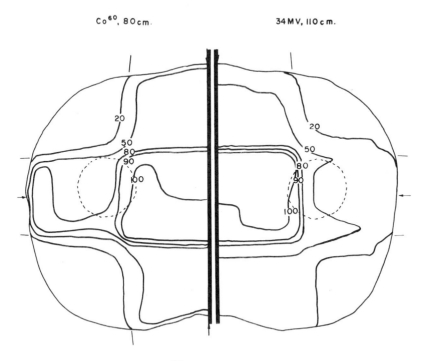

Figure 5. Comparison of radiation by ^{60}Co radiation and 34-MV photons, three-field plan.

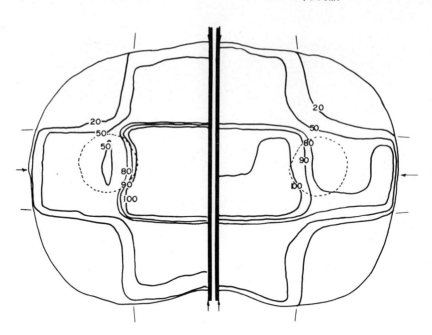

Figure 6. Comparison of radiation by 34-MV photons and by 6-MV photons in a patient 25 cm thick.

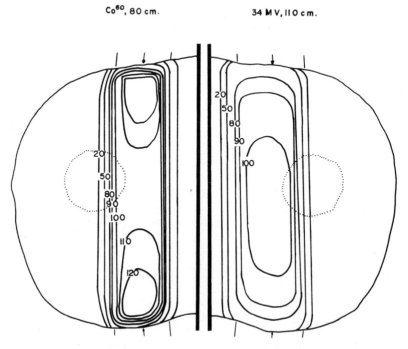

Figure 7. Comparison of radiation by 34-MV photons and by ^{60}Co radiation in a patient 25 cm thick.

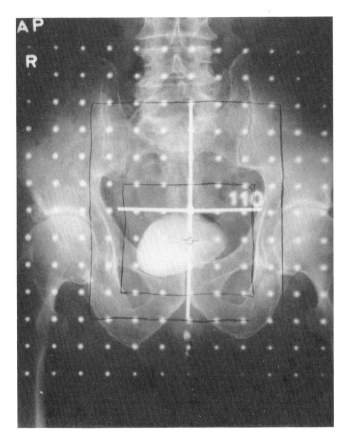

Figure 8. Front view of pelvis of patient with bladder cancer to be treated with very high-energy photons.

When there has been a previous abdominal operation, e.g., in those in whom irradiation is considered necessary following a hysterectomy for carcinoma of the cervix, a somewhat lower dose is prescribed—5,000 rads maximum in five weeks in 20 treatments—the treatment being continued beyond this time if the patient tolerates the course reasonably well. It is felt that when there is a possibility of fixation of some of the loops of the small bowel in the pelvis, a lower dose must be planned originally because of the severity of small-bowel reaction.

In obese patients suffering from early cancer of the cervix, in whom the primary treatment is by intracavitary radium supplemented by external irradiation to the lateral pelvic walls, the external irradiation, using parallel, opposing "split" fields, can be much more efficiently and elegantly delivered by very high-energy photons. Figures 6 and 7 show a considerable advantage of 34-MeV photons over those of both 6-MeV and cobalt-60, in a patient 25-cm thick.

Most of the patients with cancer of the bladder treated with this modality are suffering from Stages T_3 and T_4, and in these cases, the whole pelvis is irradiated to a maximum

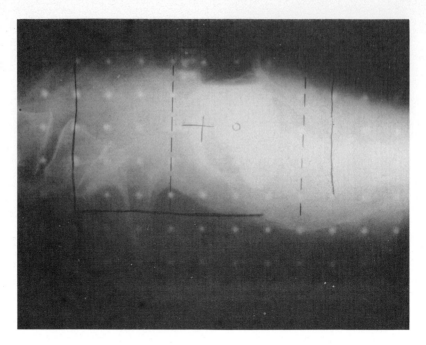

Figure 9. Side view of patient shown in Figure 8.

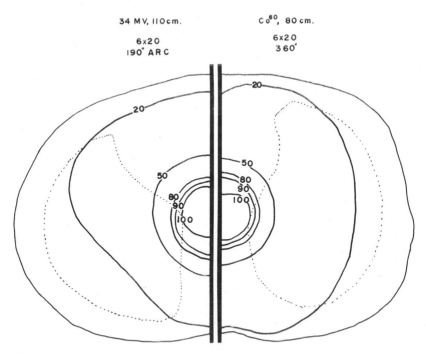

Figure 10. Comparison of similar size rotating fields using ^{60}Co radiation and 34-MeV photons.

dose of 4000 rads (minimum, 3600 rads) in 16 treatments over 4 weeks, using a three-field technique similar to that for cervix. This is followed by a 2-week course of eight treatments of 2000 rads maximum, using rotation, the latter high-dose volume being approximately an 8-cm-diameter cylinder of 10-cm length (Figs. 8 and 9). Cancer of the rectum is treated by a technique similar to that of cancer of the cervix with individualized field sizes. Techniques similar to these in the pelvis are also applicable to locally recurrent carcinoma of the endometrium and cancer of the ovary.

It is convenient to use our particular betatron for irradiation of carcinoma of the esophagus. With the patient supine and with simple simulation, a very homogeneous distribution is obtainable using 190° rotation. The actual center of the rotation is positioned a little posterior to the actual desired center of the treated volume, as one would arrange for 360° rotation, but the distribution is even and satisfactory. A comparison of similar size fields and conditions using cobalt-60 irradiation is made in Figure 10 with those produced with 34-MeV photons.

There are numerous situations of miscellaneous nature in which these high-energy photons can be usefully employed. For instance, some lesions around the head and neck can be efficiently irradiated, sometimes with a single field directed from the contralateral side. There is considerable skin and normal-tissue sparing on the uninvolved side, and under special circumstances this can be a logical choice. Pretreatment of recurrent, deeply seated tumors, through previously irradiated through overlying skin and soft tissue, similarly has obvious advantages. In many other sites, e.g., bronchogenic carcinoma, while there does not seem to be any real theoretical or practical advantage in the use of these photons, they can be used in a routine way as an alternate to lower-megavoltage equipment. Various combinations of high-energy photons and electrons are advantageous under certain circumstances. While 34-MeV photon therapy, in our experience, seems to offer practical advantages in the treatment of a rather limited number of deep-seated tumors, it is only in a few instances that marked superiority is considered to be present, and the enhancement of therapeutic effect is probably due entirely to the more favorable distribution of high-dose volumes in deep tissues, rather than in other kinds of special biological effect.

REFERENCES

Allt, W.E.C. 1969. Supervoltage radiation treatment in advanced cancer of uterine cervix: A preliminary report. *Canad Med Assoc J* 100:792–797.

Chase, H.B., H. Quastler & L.S. Skaggs. 1947. Biological evaluation of 20 million volt roentgen rays. II. Decoloration of hair in mice. *Amer J Roentgenol* 57:359–361.

Fletcher, G.H. 1964. Radiation therapy for cancer of the uterine cervix. *Postgrad Med* 35:134–142.

Gray, L.H. 1947. The distribution of the ions resulting from the irradiation of living cells. Report of a London conference held May 13–14, 1946 on certain aspects of the action of radiation on living cells. *Brit J Radiol Suppl* 1:7–15.

Harrington, E.L., R.N.H. Haslam, H.E. Johns & L. Katz. 1949. The betatron building and installation at the University of Saskatchewan. *Brit J Radiol* 110:283–285.

Horsley, R.J., H.E. Johns & R.N.H. Haslam. 1953. Energy absorption in human tissue by nuclear processes with high-energy x-rays. *Nucleonics* 11:28–31.

Johns, H.E., E.K. Darby & R.O. Kornelsen. 1951. Radiotherapeutic physics. A symposium of papers read at the 6th International Congress of Radiologists, London, July 1950. 1. The physical aspects of treatment of cancer by 22 MeV x-rays. *Brit J Radiol* 24:355–364.

Kerst, D.W. 1943. The betatron. *Radiology* 40:115–119.

Podgorsak, E.B., J.A. Rawlinson, H.E. Johns. 1975. X-ray depth doses from linear accelerators in the energy range from 10 to 32 MeV. *Amer J Roentgenol* 123:182:–191.

Quastler, H., & R.K. Clark. 1945. Biological evaluation of 20 million volt roentgen rays. 1. Acute roentgen death in mice. *Amer J Roentgenol* 54:723–727.

Quastler, H., G.D. Adams, G.M. Almy, S.M. Dancoff, A.O. Hanson, D.W. Kerst, H.W. Koch, L.H. Lanzl, J.S. Laughlin, D.E. Riesen, S. Robinson, Jr., V.T. Austin, T.G. Kerley, E.F. Lanzl, G.Y. McClure, E.A. Thompson & L.S. Skaggs. 1949. Techniques for application of the betatron to medical therapy with report of one case. *Amer J Roentgenol* 61:591–625.

Stone, R.S. 1948. Neutron therapy and specific ionization. *Amer J Roentgenol* 59:771–785.

Watson, T.A. & C.C. Burkell. 1951. The betatron in cancer therapy. Part I. *J Canad Assoc Radiol* 2:60–64.

Watson, T.A. & C.C. Burkell. 1952. The betatron in cancer therapy. Part II. *J Canad Assoc Radiol* 3:25–28.

Watson, T.A., H.E. Johns & C.C. Burkell. 1954. The Saskatchewan 1,000 Curie cobalt 60 unit. *Radiology* 62:165–176.

Watson, T.A. & C.C. Burkell. 1959. Five-years results of betatron x-ray therapy. *Brit J Radiol* 32:143–151.

Current Techniques in the Use of Nonstandard Fractionation

Jean Dutreix, M.D.,
Professor of Radiotherapy,
Institût Gustave Roussy,
Villejuif, France

The therapeutic merit of fractionation became evident in the '30's and has remained undisputed for the majority of therapeutic problems. However, the usual schedule of five fractions a week is more a social convenience than a scientific choice. The fraction size has been adjusted to approximately 200 rads on this fraction rhythm, on the basis of clinical use. Satisfactory results are obtained with this standard fractionation in many clinical situations. However, one may wonder whether a universal schedule can meet the diversity of the biological features of therapeutic problems and whether an inappropriate time-dose distribution is not responsible for some failures of radiation therapy.

Radiobiology has shed some light on the role of time factors and on the differential effect which can be related to these factors. Reducing the size of the fractions should cause a relative protection of the cell population, with the broader shoulder of the cell survival curve. Increasing the overall time of the irradiation by a larger interval between fractions, or by a gap in the treatment course, should favor the cell population with the faster repopulation rate. Fraction size and spacing between the fractions should also play a role in the reoxygenation of hypoxic cells and eventually in the kinetics of the cell cycle.

Thus radiobiology has identified the biological mechanisms related to the time-dose factors, and it suggests that for each clinical situation there should be an optimal schedule. The usual daily fraction size of 200 rads may be close to the optimum for most clinical problems; however, it is unlikely to be the best for all of them.

An example is given by some melanomas, for which large doses appear much more efficient than small doses. An explanation is provided by some experimental data (Malaise et al., 1975) showing that the initial shape of the cell survival curve may be significantly different for different cell lines; some of them obviously require large fractions for a significant cell depletion, and fractions of 200 rads appear to be almost inefficient. Differences in the shape of the cell survival curves have also been found for normal tissues (Dutreix, 1974), and a better relative protection of some tissues (for instance, lung) could be achieved with smaller fraction doses.

However, the radiobiological parameters needed for a rational choice of the optimal treatment schedule are usually missing. Thus the clinical investigation remains, at the moment, the best way of exploring the eventual therapeutic advantage of a modification of the usual time-dose distribution.

Most of the clinical investigations have been directed towards an increase of the fraction size and a reduction of the fraction number; a drastic reduction of the fraction number leads to concentrated irradiation, or even to single-dose irradiation. Fewer investigations have been done in the opposite direction of multifractionation with several fractions a day or continuous low-dose-rate irradiation.

INCREASED FRACTION SIZE, REDUCED FRACTION NUMBER

Reduced Weekly Fractionation

Fractionation three times a week is commonly used in many centers. It was introduced at our institution in 1962, for technical reasons, without modification of the weekly dose (3 x 330 rads versus 5 x 200 rads). The reduction of the fraction number had no appreciable effect on the tumor or on the normal tissues for epithelioma of the tonsil, bronchus, esophagus and bladder which were treated in this way. However, the tolerance is somewhat reduced for the more extended fields used for cervix carcinomas and Hodgkin's disease. This was attributed to a slight increase of the biological effect with reduction of the number of weekly fractions from 5 to 3 without reduction of the weekly dose.

A clinical trial has been carried out by Marcial on different cancer sites, and particularly cervix cancer, comparing 5 X 150 rads to 3 X 250 rads per week (the same weekly dose) for 6 weeks of treatment. No difference was observed in early reactions, late complications, and 3-year survival rates.

An extensive clinical trial has been organized by 17 British centers (BIR Working Party, 1963, 1972) comparing 5 versus 3 weekly fractions for treatment of carcinoma of the larynx and pharynx, the weekly dose being decreased by 10% in the second group, according to the NSD formula (Ellis, 1967).

The September 1974 report (G. Wiernik & J.F. Fowler, private communication) concerns 532 patients: the late survival at 5-6 years is slightly higher for 3 weekly fractions (without any statistical significance); the late reactions of skin, mucous membrane, and subcutaneous tissues and the incidence of edema and myelitis are the same for both groups.

A different conclusion is drawn by Kok (1973) from the comparison of 5 X 210-230 rads with 3 X 270-300 rads per week. Less tumor control was obtained and the reaction was more severe for the second group, in which the weekly dose was reduced by 20% with respect to the first group.

Montague (1968) has reported a comparison of two subsequent series of breast cancer treated 5- and 3-times-per-week with the same weekly dose of 1000 rads. The percentage of local recurrence was smaller with the 3-times-a-week schedule, but the severity of late complications was greater and considered prohibitive.

Comparisons of 2 versus 5 weekly fractions have been reported in some studies. A trial on lung cancer was performed by Deeley (1974), comparing 2 X 400 rads to 5 X 200 rads per week: the survival rate, the incidence of metastases, and the morbidity were the same. Holsti (private communication) has compared 2 X 450 rads to 5 X 200 rads per week on lung tumors and osteosarcomas without finding evidence of any differences. Angelakis et al. (1973) has compared 2 X 280 rads to 5 X 200 rads per week (weekly dose: 760 rads versus 1000 rads) on advanced cases and metastases and observed that the biological effects are less for 2 X 380 rads per week than for 5 X 200 rads, without evidence of any differential effect between normal and tumor tissues. Table 1 summarizes the fractionation schema used in the clinical studies described above.

TABLE 1. CLINICAL STUDIES WITH REDUCED WEEKLY FRACTION NUMBER

Fractions/Week	Reduced Fractions/Week	Investigator
5 X 200 = 1000	3 X 330 = 1000	Montague
5 X 200 = 1000	3 X 300 = 900	BIR
5 X 220 = 1100	3 X 285 = 855	Kok
5 X 150 = 750	3 X 250 = 750	Marcial
5 X 200 = 1000	2 X 450 = 900	Holsti
5 X 200 = 1000	2 X 400 = 800	Deeley
5 X 200 = 1000	2 X 380 = 760	Angelakis

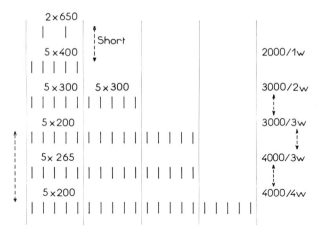

Figure 1. Example of a clinical trial designed for investigating the individual role of different factors (number of weekly fractions, fraction size, weekly dose, total dose, overall time).

In some clinical studies a reduction of the fraction number is associated to a reduction of the overall time, by keeping the usual rhythm of 5 fractions a week; Stell and Morrison (1973) have published results of treatments with 15 fractions of approximately 365 rads in 3 weeks.

Fletcher (1973) has compared treatments of 5000 rads (20 X 250 rads) in 4 weeks, 6000 (25 X 240) in 5 weeks, and 7000 rads in 6½ weeks (33 X 210 rads). In these (larynx) studies, the complications appear to be more severe with the shorter treatment, with the larger fraction size. However, the role of fraction size and of overall time cannot be distinguished.

A comprehensive protocol has been applied to brain metastases by the Radiation Therapy Oncology Group in the United States. It allows (Fig. 1) investigation of the individual roles of the different factors: number of weekly fractions, fraction size, weekly dose, total dose, overall time. The preliminary results suggest that the best schedule should be 4000 rads/3 weeks. (Kramer, personal communication). The results of the clinical studies comparing 5 fractions a week to 3 or 2 fractions a week do not allow a definitive conclusion.

For most of the clinical series, the effects on tumor, as well as on normal, tissues appear to be the same when changing from 5 to 3 fractions a week, with a reduction of about 10% of the weekly dose. However, there remain some discrepancies, and as pointed out by Fletcher (1974), particular attention should be paid to the analysis of the late injuries. The diversity of the fraction sizes which have been used should be noted.

The sensible approach to checking the possibility of a therapeutic differential effect between two different schedules consists in matching the fraction doses in order to get the same effect on the normal tissues.

This implies a choice between the isoeffect dose and the fraction number or the fraction size. The most commonly accepted relationship is the NSD (Ellis , 1967); however, other relationships lead to different correspondences between the fraction doses (Table 2; Fig. 2).

TABLE 2.

Equivalent* Weekly Dose	Equivalent Single Dose for 4 Weeks Treatment			
	N.S.D. (Ellis)	Difference**	C.P.K. (Cohen)	Difference**
5 X 200 rads	1362 rads	0	1207	0
4 X 245 rads	1408 rads	3 %	1219	1 %
3 X 315 rads	1455 rads	7 %	1266	5 %
2 X 440 rads	1493 rads	9 %	1454	20 %
1 X 750 rads	1503 rads	10 %	1886	56 %

* Equivalent doses computed from data on human skin (Dutreix & Wambersie, 1973).
** Difference with respect to single dose equivalent to 5 X 200 rads weekly dose.

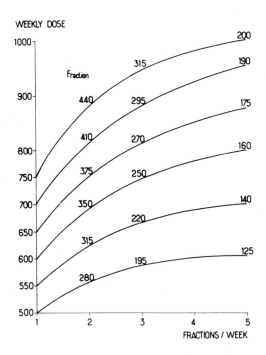

Figure 2. Equivalent weekly dose for different numbers of weekly fractions [values computed from clinical data obtained on human skin desquamation—(Dutreix and Wambersie, 1973)].

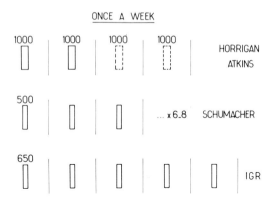

Figure 3. Examples of once a week fractionation.

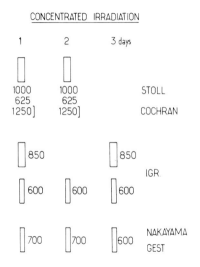

Figure 4. Examples of concentrated irradiations.

More numerous and more reliable data on the isoeffect dose as a function of the fraction size, for the early and late effects on the normal tissues involved in the irradiation, would be very helpful in the design of a comparative study and in the analysis of its results.

Once-a-week fractionation

This kind of fractionation, departing drastically from standard fractionation, is used by some authors for palliative irradiation; Schumacher (1972) obtains good results on lung cancer with single weekly doses of 500 rads. It is also used in several centers for some particular tumors, such as osteosarcomas, melanomas, etc.

We have tried to use this schedule for epithelioma of the esophagus with 5 fractions of 650 rads once a week, assumed from our data (Dutreix and Wambersie, 1973) to be equivalent to 3 X 300 rads or 5 X 190 rads per week. The trial was stopped after treating five patients for the following reasons:

1) The tolerance was poor, with complaints of mediastinal pain and oesophagitis,
2) There was no relief of the dysphagia,
3) A fistula between digestive and respiratory tract was found in the three necropsies which were performed and the tumor was still present, and,
4) The survival time was 1 to 7 months, with an average of 4 months, much shorter than with more conventional schedules.

Figure 3 gives examples of once-a-week fractionation.

Concentrated Irradiation

Concentrated irradiation with a few fractions given in a few days has significant advantages in palliative and preoperative irradiation. Some examples are shown in Figure 4.

It was advocated by Cochran et al. (1959), Horrigan et al. (1962), Stoll (1964), Atkins (1964), and Edelman et al. (1975), for palliative irradiation of inoperable carcinomas of the breast, and by Levitt et al. (1965) and Abramson and Cavanaugh (1970) for lung carcinomas. It has been used by Nakayama (1964), Gary Bobo et al. (1972) for preoperative irradiation of cancer of the oesophagus, and by Bachelot et al. (1972) for preoperative irradiation of cancer of the floor of the mouth. The dose is usually given in 2 to 3 fractions in consecutive days. However, single doses of 1000 to 2000 rads are considered satisfactory by Nakayama (1964) for breast and rectum, and Hindo et al. (1970) and Shehata et al. (1974) for brain metastases. Jazy and Aron (1974) concluded that the results of the single dose were not as good as the reported results of fractionated irradiation.

A single dose of 1000 rads is compared to several fractionated irradiations in a clinical trial of the Radiation Therapy Oncology Group (F.R. Hendrickson, private communication). J. Kujawska (private communication) on skin epithelioma treated by superficial radiation therapy has compared 1 X 2250 rads to daily fractionation of 4 X 1000 rads and 8 X 600 rads. At 3 years' follow-up the recurrence rate is larger for the last group without statistical significance, and cosmetic results have not yet been evaluated.

The repetition of the concentrated irradiation is considered by Shehata (1974) for brain tumors. In a clinical trial on bladder carcinomas, Harvey et al. (1971) have compared two series of concentrated irradiation with 3 to 4 weeks' interval to 6000-7000 rads conventional fractionation and have concluded that there is some advantage to the concentrated irradiation.

Schlienger et al. (1973) have compared, on brain tumors, concentrated irradiation (2 X 850 rads repeated at 4-week intervals) to fractionated irradiation (5 to 6000 rads in 5 to 7 weeks). They concluded that the concentrated irradiation is advisable for the glioblastomas; however, the fractionated irradiation appears to be more satisfactory for the other histological types.

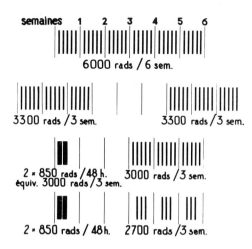

Figure 5. Split-course therapy with concentrated first series (Dutreix et al., 1966). The reduction to 2 fractions of the first series allows the same overall time as for the corresponding uninterrupted irradiation.

Levitt et al. (1965) have compared for lung cancer an intensive split-course (3 X 600 rads/28 days rest/3 X 600 rads) to a conventional 6000 rads/6 weeks irradiation. A decrease in tumor response was apparent and an increase of non-symptomatic fibroses in the patients treated with the intensive split-dose approach. However, there was no difference in survival or relief of symptoms. A change of the 3 X 600 rads sequence into 5 X 450 rads was not well-tolerated; the results with 5 X 400 rads were encouraging.

The concentrated irradiation of 2 X 850 rads at 48-hour intervals has been used extensively at Gustave Roussy Institute for palliative irradiation or as preoperative irradiation (Dutreix et al., 1966, 1971). The tolerance appears to be acceptable in most cases; however, it has been replaced by 3 X 600 rads in 3 days for whole brain irradiation, with a significant reduction of headache and nausea, and also for extended fields on the thorax and pelvis.

The concentrated irradiation was designed at Gustave Roussy Institute not only as a palliative irradiation, but as the first part of a treatment with a curative intent (Fig. 5). The basic idea was that by concentrating the first half of the treatment into 2 fractions, a faster effect on the tumor would be obtained and a rest interval could be provided without an increase in the overall time of irradiation. The second series is given 3 weeks later as a fractionated irradiation. The biological advantages expected from the rest interval have been discussed by the promoters of the split-course technique. In addition, the short first series can be used as preoperative irradiation without delay in the surgery.

A clinical trial (Dutreix et al., 1974a,b) has been carried out on epithelioma of the tonsil, comparing the split-course technique with a concentrated first series with uninterrupted fractionated irradiation. The tumor regression and the mucositis occurred earlier with the split-course technique. The degree of early and late reaction of the oral mucosa and of the skin were the same for both techniques, and there was no difference in 5-year survival rate. Thus these results did not show any therapeutic superiority for any of the techniques which were compared; however, they demonstrated that at least for this tumor site, the first 3 weeks of the fractionated irradiation could be reduced to 2 fractions without any risk to the patient.

The split-course technique with a concentrated first series is the usual treatment schedule at Gustave Roussy Institute for carinomas of the bladder, lung, and oesophagus (Fig. 5). A question arose from the clinical findings for cancer of the esophagus that the first series (2 X 800 rads) causes a significant improvement in dysphagia, radiological symptoms, weight and general condition (Karnofsky index), while the second series (9 X 300 rads/3 weeks) after 3 weeks rest does not add very much to this improvement. The question was whether a longer and more comfortable survival could not be obtained by delaying the second series.

A clinical trial (Dutreix et al., 1974b) has been conducted in which systematic irradiation (3 weeks rest between the two series) was compared with a symptomatic irradiation, i.e., one in which the second series was given only when necessary because of the symptomatology. The conclusion of the trial is that the survival curve is the same for the two protocols up to 3 years, including the number of gastrostomies. However, the relief of the major symptoms is greater and longer with the systematic irradiation and thus a more comfortable survival is attained.

MULTIFRACTIONATION

Irradiation with 2 fractions per day has been used at many centers (Table 3). Fletcher et al., (1974) have compared 40 daily fractions (of 150 rads) to the same number of fractions given twice a day, the overall time being reduced to half and the fraction size being reduced by 10%. This study is actually a comparison of two different periods of

TABLE 3. MULTIFRACTIONATION

Daily Dose (Rads)	Interval (Hours)	Total Dose (Rads)	Investigator
2 X 150	8	6000	Fletcher
1 X 150		6000	
2 X 190	2	5300	IGR
1 X 330		4600	
3 X 50	8	3000	
65	8	to	Simpson
145	8	4000	
3 X 100	8	8400	Littbrand
1 X 200		6400	

overall times and a greater effect on the tumor (inflammatory breast cancer) and on normal tissues was found with the shorter course.

A clinical trial has been carried out at the Gustave Roussy Institute comparing a conventional fractionation treatment (A, 3 X 330 rads/week) to treatment B, in which each 330 rads fraction is replaced by 2 subfractions of 190 rads at 2-hour intervals. The aim of this trial was to compare 3 with 6 weekly fractions. The short interval of 2 hours between the subfractions, has been adopted as sufficient to approach a full cellular repair of sublethal injuries for skin and mucosa; it could possibly bring some additional differential effect if the cellular repair were slower for the tumor.

The equivalence 330 $\rightleftharpoons$ 2 X 190 rads was adopted on the basis of the clinical experience of some authors, suggesting that the weekly dose should be increased by 15% when changing from 3 to 5-6 fractions per week. This clinical trial has been made on epithelioma of the tonsil, and 30 patients have been entered in each group. The immediate observations have shown that treatment B has a greater effect on reactions of skin and mucosa, as well as tumor regression. The conclusion is that 2 X 190 rads at 2-hour intervals is equivalent, for these early effects, to more than 330 rads, and from a subsequent study (Dutreix, 1973) the correct correspondence was judged to be 330 rads $\rightleftharpoons$ 2 X 175 rads. Thus treatment B corresponds to a small overdosage with respect to A.

The disappointing fact is that the 5-year survival rate is significantly less for treatment B than for treatment A. No explanation for this poorer result has been found in late complications on normal tissues; the late effects on normal tissues are moderate and similar for both treatments. Thus treatment B, which was more efficient than treatment A on the bulk of the tumor, as shown by the faster initial regression, has had a lower efficiency for the part of the tumor responsible for the final survival. The total number of fractions does not seem responsible for this failure, since the results published in the literature for the conventional 5 to 6 weekly fractionations are better than those obtained with treatment B (while they are similar to the results of treatment A). The weekly dose and total dose for treatment B keep within the limits of standard treatments.

The interval of 2 hours between the subfractions may be responsible for this failure, possibly through unfavorable synchronization. However, ethical restrictions raised by the late results of this trial have discouraged further clinical trials to study different intervals between the subfractions.

Simpson (1969) has carried out an admirable clinical trial on glioblastomas, comparing 3 fractions a day at 8-hour intervals to daily fractions, with the same total dose of 3000 to 4000 rads, given over 7, 21 or 28 days. No difference was observed in the length of remission or the survival rate.

Littbrand et al. (in press), for carcinoma of the bladder, have compared a daily fraction of 200 rads to 3 X 100 rads per day. The irradiation is given 5 times a week, in 2 months, with a rest period of 2 weeks in the middle of the treatment (total doses, 6400 and 8400 rads). The results obtained on the first 45 patients of the clinical trial suggest an improved therapeutic effect with the multifractionated irradiation.

Continuous irradiation is advocated by Pierquin et al. (1974), who have obtained lasting control of advanced tumors with continuous daily irradiation, approximately 1000 rads in 8 hours for 3-6 consecutive days. A modification of the technique consists of a split-course irradiation, with two series of 3500 rads in three consecutive days, separated by a rest interval of 2 weeks.

CONCLUSION

Numerous clinical investigations and trials have been carried out to explore the therapeutic advantage of deviations from conventional fractionation. The clinical studies reported in this paper represent illustrative examples rather than an exhaustive survey. Most of these studies have failed to demonstrate the therapeutic advantage anticipated. A pessimistic view is that the time-dose distribution is not critical. This is likely to be true for many clinical situations, for which the irradiation only results in a palliative effect, with a very loose correspondence between the therapeutic effect and the total dose. On the contrary, when the therapeutic effect is sharply dependent on the dose, one may expect an improved therapeutic ratio from a slight additional difference between the biological effectiveness on tumor and normal tissues. The possibility of bringing about such a favorable differential effect by altering the fraction size, or the overall time of the irradiation, or both, should be further tested with trials on the time-distribution.

REFERENCES

Angelakis, P., C., Papavasiliou & C. Elias. 1973. Twice per week treatment versus five times per week. A radiotherapeutic clinical trial. *Brit J Radiol* 46:350–353.

Abramson, N. & P.J. Cavanaugh. 1970. Short-course radiation therapy in carcinoma of the lung. *Radiology* 96:627–630.

Atkins, H.L. 1964. Massive dose technique in radiation therapy in inoperable cancer of the breast. *Amer J Roentgenol* 91:80–89.

Bachelot, F., P. Guerin, G. Delouche & J. Gest. 1972. Cancer du plancher de la bouche. Analyse de 70 cas. *Semaine des Hôpitaux* 48:1413–1423.

Cochran, D.Q., S. Holtz & W.E. Powers. 1959. The rapid palliative treatment of breast carcinoma. A preliminary report. *Amer J Roentgenol* 81:479–484.

Cohen, L. 1968. Theoretical "isosurvival" formulae for fractionated radiation therapy. *Brit J Radiol* 41:522–528.

Cohen, L. 1973. An interactive program for standardization of prescriptions in radiation therapy. *Computer Progr Biomed* 3:27–35.

Deeley, T.J. 1974. Radiotherapy of carcinoma of the bronchus. *Cancer Treatment Reviews* 1:39–64.

Dutreix, J., M. Schlienger & R. Couvelaire. 1966. La radiothérapie pré-opératoire en 2 séances. *J Urol Néphrol* 72:364–366.

Dutreix, J., M. Schlienger, C. Chauvel & R. Daguin. 1971. Concentrated palliative radiotherapy for tumors affecting the esophagus, brain, bones and mediastinum. *Ann Clin Res* 3:9.

Dutreix, J. & A. Wambersie. 1973. Relation between total dose and number of sessions for reaction of the human skin. Abstract No. 192, p. 71. *In* Radiology, Proc. XIII Intl. Congr. Radiol., Madrid, October, 1973. International Congress Series No. 301. Excerpta Medica, Amsterdam.

Dutreix, J., M. Hayem, B. Pierquin, K. Zummer, C. Hesse & A. Wambersie. 1974a. Epithéliomas de la région amygdalienne. Comparaison entre fractionnement classique et irradiation en deux séries (split-course). *Acta Radiol* 13:167–184.

Dutreix, J., J. Robillard, K. Rubinstein-Zummer, A. Roussel & A. Chavy. 1974b. Essai thérapeutique sur l'irradiation systématique et l'irradiation symptomatique des épithéliomas de l'oesophage. *Bull Cancer* 61:265.

Dutreix, J. & A. Wambersie. 1974c. Iso-effect of total dose as a function of the number of fractions for skin, intestinal mucosa and lung. In Caldwell, W.L. and D.O. Tolbert (eds.) Proc.Time-Dose Conf., Madison, October, 1974. Univ. of Wisconsin Press.

Edelman, A.H., S. Holtz & W.E. Powers. 1965. Rapid radiotherapy for inoperable carcinoma of the breast. Benefit and complications. *Amer J Roentgenol* 93:585–589.

Ellis, F. 1967. Fractionation in radiotherapy. p. 34–51. In Deeley, T. & C. Wood (eds.) Modern Trends in Radiotherapy. Butterworths, London.

Fletcher, G.H. 1973. Textbook of Radiotherapy. 2nd edition. Lea & Febiger Publishers, Philadelphia.

Fletcher, G.H. & H.T. Barkley. 1974. Present status of the time factor in clinical radiotherapy. I. The historical background of the recovery exponents. *J Radiol Electrol* 55:443–450.

Fletcher, G.H., H.T. Barkley & L.J. Shukovsky. 1974. Present status of the time factor in clinical radiotherapy. II. The nominal standard dose formula. *J Radiol Electrol* 55:11, 745–541.

Gary-Bobo, J., E. Negre, H. Pujol, J.C. Laurent, R. Ramos & J. Bonaccorsi. 1972. L'irradiation concentrée pré-opératoire. Apport dans le traitement radio-chirurgical du cancer oesophagien. A propos de 110 cas. *J Electrol Med Nucl* 53:893–894.

Gest, J. & G. Delouche. 1966. La radiothérapie concentrée. Résultats immediats et possibilitiés. *J Radiol Electrol* 47:812–817.

Harvey, P., M. Brunet, F. Bachelot & J. Gest. 1971. Résultat du traitement du cancer de la vessie par cobalthérapie concentrée. (Etude de deux séries comparatives) *Ann Urol* 5:183–188.

Hindo, W.A., F&A. DeTrana, M.S. Lee et al. 1970. Large dose increment irradiation in treatment of cerebral metastases. *Cancer* 26:138–141.

Horrigan, W.D., H.L. Atkins & N. Tapley. 1962. Massive-dose rapid palliative radiotherapy. *Radiology* 78:439–444.

Jazy, F. & B.S. Aron. 1974. Single dose irradiation in treatment of cerebral metastases from bronchial carcinoma. *Cancer* 34:254–256.

Kok, G. 1973. NSD for treatment of carcinoma of the larynx (abstract). *Int J Radiat Biol* 24:317.

Levitt, S.H., P.T. Condit & C.R. Bogardus, Jr. 1965. Split-dose intensive radiation therapy in the treatment of patients with advanced carcinoma of the lung. *Radiology* 85:738–779.

Levitt, S.H., T.K. Jones, S.J. Kilpatrick & C.R. Bogardus, Jr. 1969. Treatment of malignant superior vena cava obstruction. A randomized study. *Cancer* 24:447–451.

Littbrand, B., F. Edsmyr & L. Revesz. 1975. A low dose fractionation scheme for the radiotherapy of carcinoma of the bladder. *Bull Cancer* 62:241–248.

Malaise, E.P., J. Weininger, A.M. Joly & M. Guichard. 1975. Measurements in vitro with three cell lines derived from melanomas. p. 223. *In* Alper, T. (ed.) Proc. Sixth L. H. Gray Conference, 1974. J. Wiley & Sons.

Marcial, V.A. 1969. Studies on the relationships between dose-time, fractionation, and rest period in radiation therapy. p. 280–285. *In* Time and Dose Relationships in Radiation Biology as Applied to Radiotherapy, Proc. NCI-AEC Conf. Carmel, California, September, 1969. Brookhaven National Laboratory.

Montague, E.D. 1968. Experience with altered fractionation in radiation therapy of breast cancer. *Radiology* 90:962–966.

Nakayama, K. 1964. Pre-operative irradiation in the treatment of patients with carcinoma of the oesophagus and of some other sites. *Clin Radiol* 15:232–241.

Pierquin, B., F. Baillet & C. Brown. 1974. La téléradiotherapie continue et de faible débit. Deuxième rapport. *J Radiol Electrol* 55:757–763.

Schlienger, M., J.P. Constans, J. Roujeau, S. Askeinazy & F. Eschwege. 1973. Irradiation d'une série de 304 tumeurs intracrâniennes malignes primitives de l'adulte. *J Radiol Electrol* 54:939–950.

Schumacher, W. 1972. Nutzbarmachung neuer Erkenntnisse über die Fraktionierung bei der Bestrahlung bosartiger Tumoren für die Praxis. *Rontgen-Berichte* 1:92.

Shehata, W.M., F.R. Hendrickson & W.A. Hindo. 1974. Rapid fractionation technique and re-treatment of cerebral metastases by irradiation. *Cancer* 34:257–261.

Simpson, W.J. 1969. Discussion. p. 301. *In* Time and Dose Relationships in Radiation Biology as Applied to Radiotherapy, Proc. NCI-AEC Conf. Carmel, California, September 1969. Brookhaven National Laboratory.

Stell, P.M. & M.D. Morrison. 1973. Radiation necrosis of the larynx. *Arch Otolaryngol* 98:111–113.

Stoll, B.A. 1964. Rapid palliative irradiation of inoperable breast cancer. *Clin Radiol* 15:175–178.

Working Party of the British Institute of Radiology. 1963. Preliminary report on the effects of dose fractionation in radiotherapy. *Brit J Radiol* 36:382–383.

Working Party of the British Institute of Radiology. 1972. Fifth progress report of the fractionation trial 3F/week or 5F/week treatment of larynx and pharynx. *Brit J Radiol* 45:754–756.

Dosimetry Considerations
of Electron Beams

Peter R. Almond, Ph.D.,
Department of Physics,
The University of Texas System Cancer Center,
M.D. Anderson Hospital and Tumor Institute,
Houston, Texas

The purpose of this presentation is to provide some basic data and information for practical electron-beam dosimetry. It will include a discussion on the basic parameters of electron beams, measurement of absorbed dose, beam control and beam characteristics. Some suggestions for practical, routine dosimetry and treatment-planning will be given.

Electrons with energies from 1-50 MeV will be considered, with most of the data being for electrons of 6-30 MeV. If lower-energy electrons are used (to treat *mycosis fungoides*, for example), special procedures are required, as described in the literature (Karzmark, 1968). For the higher energies the data presented here can successfully be extrapolated up to 50 MeV.

BASIC PARAMETERS OF ELECTRON BEAMS

Since the single characteristic of the electron beam which has provided the basis for its use in radiotherapy is the sharp fall-off in depth-dose, and the depth of this fall-off is a function of the energy, it is necessary to define and characterize the electron-beam energy. Parameters used for dosimetry calculations are also energy-dependent. Some of these parameters may also depend upon the energy spectrum of the electron beam, which describes the quality of the electron radiation. The shape of the fall-off for the depth-dose curve is also dependent upon the quality of the beam, which is affected by collimation and scattering of the beam through flattening foils.

ELECTRON BEAM ENERGY

Figure 1 shows the various energies that need to be defined. E_a is the energy to which the machine accelerates the electrons, and in general this beam has a very narrow energy spread. E_o is the energy at the body surface, after the beam has passed through the accelerator window, scattering foils and air layers. Energy losses of the electrons in these layers of matter shift the electron spectrum to lower energies and broaden the electron spectrum. Provided the energy losses are relatively small as compared to E_a, a single energy value will be considered at the surface and

$$E_o = E_a - \Delta E_{tot} \qquad (1)$$

where ΔE_{tot} is the total energy loss in going through the window, foils and air column to the surface. The contribution to ΔE_{tot} from each single layer of thickness dl and stopping power S is $dl.S$ (see below). In characterizing an electron beam by a single energy parameter, it is recommended that the value of E_o be used after the beam has passed through all scattering foils, etc., that are normally used.

ENERGY SPECTRA

As noted above, the quality of the electron radiation is characterized by its energy spectrum, which will affect the various energy-dependent parameters. Contributions to the spectrum are made by: primary electrons, secondary electrons, and electrons released

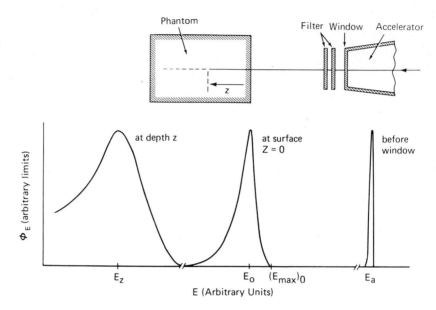

Figure 1. Definition of various energies used in electron-beam physics. ϕ_E is the distribution of fluence in energy and $(E_{max})_0$ is the maximum energy of the electrons at the surface. (Redrawn from ICRU, 1972.)

by the reabsorption of bremsstrahlung produced within the medium (only significant at the higher energies).

The spectrum at the entrance surface of an absorber is characteristic of the accelerator and the various scattering foils and collimators, etc. The introduction of the scattering foils will broaden the energy spectrum of the electrons and also introduce some low-energy electrons into the beam. The spectrum at the exit surface of an absorbing medium, in comparison with the spectrum at the entrance, will indicate the energy loss of the beam in penetrating the absorber and the contribution from secondary electrons (Harder, 1964). Figure 2 shows the results of such measurements. As the electrons penetrate the medium, the energy spectrum broadens; Harder has shown that the most probable energy of the electrons (defined by the position of the peak in the spectrum) decreases linearly with absorber thickness, and the mean energy of the spectrum E_Z also decreases approximately linearly (Harder, 1964). This can be expressed by the relationship

$$E_Z = E_o \left(1 - \frac{Z}{R_p}\right) \qquad (2)$$

E_o is the initial energy, R_p the practical range (described below), and Z the absorber thickness. Detailed studies, both experimental and calculational, on the contribution of the secondary electrons have shown that the low-energy component changes relatively little with depth (Berger and Seltzer, 1969).

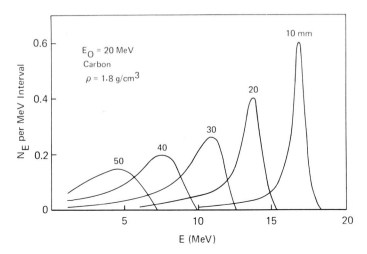

Figure 2. Energy distribution of the electron number N_E, as a function of electron energy E, behind carbon layers of different thickness d, for initial energy $E_0=20$ MeV. (Redrawn from ICRU, 1972.)

MEASUREMENT OF BEAM ENERGY

For the characterization of the electron beam energy by a single energy parameter, three methods of energy measurement can be used: nuclear threshold energy measurements, range measurements, and the measurement of the Cerenkov radiation threshold.

The nuclear threshold method requires that the electron-beam energy be varied continuously and that a Geiger-Mueller counter or scintillation counter be available. The Cerenkov radiation threshold method requires special equipment. Both systems have been described in the literature, and where comparisons have been made, they yield the same energy as that obtained by range measurements (de Almeida and Almond, 1974). To obtain the energy from the range, the well-established range-energy relationship is used. This can be stated as:

$$\rho R_p = k_1 E_0 - k_2 \qquad (3)$$

where ρ is the density of the medium in grams per cubic centimeter. R_p is the practical extrapolated range in centimeters, as defined in Figure 3.

For water, $k_1 = 0.521$ g cm^{-2} MeV^{-1}, and $k_2 = 0.376$ g cm^{-2}.

Factors that affect the measurement of the practical range include beam diameter, type and size of detector, and beam divergence. Figure 4 shows two arrangements that can be used. S is the distance from the surface of the absorber to the virtual electron beam source and Z is the depth in the absorber. It is usual to set up this measurement at the normal treatment distance, with the scattering foils and monitors in place. This will then determine the energy at the surface of the patient. Some authors have used slight variations in k_1 and k_2, but unless there is a large energy spread, equation 3 is adequate, and the energy determined may be regarded as the single valued surface energy. The accelerator energy can then be determined from equation 1.

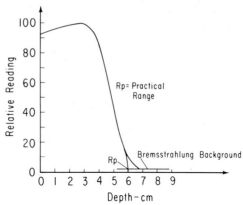

Figure 3. Definition of the practical range of electrons. The relative ionization-chamber reading is plotted versus the depth in the absorber. Also shown is the bremsstrahlung background extending beyond the range of the electrons.

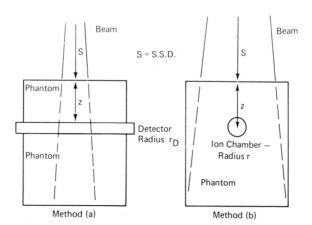

Figure 4. Methods of measuring the practical range. (a) The detector radius $r_D > \frac{1}{2} R_p$ and is also greater than the beam radius—no corrections need to be made to the ion chamber readings. (b) The detector radius is very small compared to the beam radius and the beam radius $r_B > \frac{1}{2} R_p$. The position of the center of the detector must be corrected for displacement (0.75 radius) and the ion chamber reading for beam divergence

$$J \text{ (true ion current)} = J\text{(measured ion current)} \frac{(S+Z)^2}{S^2}, \text{ where } S=\text{S.S.D.}$$

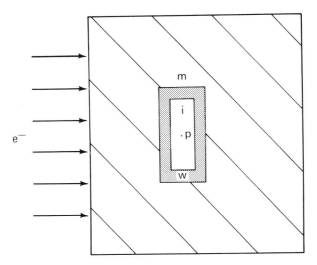

Figure 5. Diagrammatic presentation of the probe method for measuring absorbed dose. The radiation-sensitive dosimeter, i, with a wall, w, is placed at the point of interest, P, within the irradiated medium, m. (Redrawn from ICRU, 1972.)

DETERMINATION OF ABSORBED DOSE

General Concepts

Figure 5 shows the general arrangement for measuring absorbed dose, D_m, given to the medium at a point P. A small piece of the medium is replaced by the radiation-sensitive dosimeter material. The size of the detector must be small enough to minimize perturbations of the particle fluence but large enough to contain many interactions and yield a good signal. These concepts apply to all kinds of dosimeters, e.g., calorimeters, ion chambers, and chemical, photographic and thermoluminescent dosimeters.

The absorbed dose in the dosimeter D_i is measured, and D_m is calculated by the Bragg-Gray relation.

$$D_m = D_i \, S_{mi} \, p_{mi} \qquad (4)$$

where S_{mi} is the ratio of the mass stopping powers in the medium to that in the dosimeter and p_{mi} is the perturbation factor to correct for disturbances in the particle fluence. In the methods described for determining the absorbed dose, the calculation of S_{mi} will be as simple as possible, and experimental conditions will be recommended to ensure that p_{mi} deviates only slightly from unity. Because the presence of the dosimeter within the medium causes a transition of the secondary electron spectrum near the dosimeter boundary and because about 25% of the total absorbed dose is produced by these secondary electrons, the boundary effect should be suppressed. This can be done by matching the atomic composition of the sensitive volume, wall, and surrounding medium. In this case the change in the secondary electron spectrum near the boundary will be small, and the measured absorbed dose in the dosimeter will be nearly equal to the dose to the medium at the point of interest. Such will be the case for the ferrous sulfate dosimeter in a water phantom. If the sensitive material of the probe cannot be matched to the medium, then the atomic composition of the wall should be made to match the atomic composition of the sensitive volume of the dosimeter. With the correct choice of wall thickness, the boundary effects will occur within the wall. Various investigations have been made of this effect, and for electrons from 5-50 MeV, the appropriate wall thickness is from 1-2 mg/cm^2.

By assuming that one or the other of the above conditions is met so that there is complete secondary electron equilibrium, and assuming a linear relationship between $(S/\rho)_{col}$, E_z, M and the electron energy in the peak region of the primary electron spectrum, the stopping power ratio can be calculated by:

$$S_{mi} = \frac{(S/\rho)_{col}, E_z, M}{(S/\rho)_{col}, E_z, i} \qquad (5)$$

where $(S/\rho)_{col}$ has to be calculated for the mean energy E_z of the primary electrons at the point of interest. E_z can be calculated by using equation 2 and values of $(S/\rho)_{col}$ taken from Berger and Seltzer, 1966 (and NASA SP-3012). For practical purposes these equations are quite sufficient, even for the case of an air-filled probe in water. The deviation between the calculated stopping power ratio and experimental values does not exceed 2%.

Ion Chambers

By the use of a gas-filled ionization chamber, the absorbed dose D_m in the material of interest can be obtained from the absorbed dose in the gas D_i by the Bragg-Gray relations.

$$D_m = S_{mi} D_i P_{mi} = S_{mi} \frac{\overline{W}}{e} J_g P_{mi} \qquad (6)$$

J_g is the measured ionization per unit mass of the gas. W is the average energy expended in the gas per ion pair formed, e is the charge of the electron, P_{mi} is the electron fluence perturbation factor, and S_{mi} is the stopping power ratio calculated according to equation 5. When J_g is expressed in coulombs (Q) per kilogram (M), W in Joules, e in coulombs, then the absorbed dose is given by

$$D_m = S_{mi} \frac{\overline{W}}{e} \frac{Q}{M} P_{mi}* \qquad (7)$$

Since it is now usual to measure the collected ionization directly in coulombs with a calibrated electrometer, the unknown in the above equation is M. M can be determined if the chamber can be calibrated in a known radiation field in terms of exposure (usually with ^{60}Co γ-rays or 2 MV x-rays) to yield a calibration factor N. N relates the chamber reading R to the exposure in roentgens at the position of the center of the chamber when the chamber is removed. Assuming that the build-up cap, which provides electronic equilibrium, acts as an air-equivalent material for the calibration radiation,

---

* This yields the absorbed dose in Joules/kilogram. To obtain the dose in rads, the result must be multiplied by 100 since 1 rad = 0.01 Joules/kilogram.

$$M = \frac{100\ (\overline{\frac{W}{e}})\ k}{0.869\ N.A.}\ kilogram \quad (8)$$

k is a constant that relates chamber reading to charge in coulombs (i.e., $Q=kR$ and $k=1$, when the electrometer reads directly in coulombs). A is the attenuation factor for the radiation through the build-up cap at the time of the exposure calibration for ionization chamber ($A=0.985$ for ^{60}Co γ-rays). Now consider the calibration of an electron beam of energy E. The build-up cap is removed, and the chamber placed in the medium. From equation 7 and 8:

$$D_m(E) = \frac{100\ R(E)k\ (\overline{\frac{W}{e}})\ (S_{mi})}{100\ (\overline{\frac{W}{e}})\ k\ /\ N.A.0.869}\ p_{mi} \quad (9)$$

$$= R(E)\ N.A.\ (S_{mi})\ E\ 0.869\ p_{mi}$$

$$= R(E)\ N.\ C_E\ p_{mi}$$

where $R(E)$ is the chamber reading in the electron beam of energy E.

$$C_E = A.(S_{mi})E\ 0.869 \quad (10)$$

using the value of A given above and equation 5 to calculate S_{mi}, the values for C_E can be determined. Since N is a constant for a given chamber and C_E can be calculated, the chamber can be used to calibrate electron beams at any energy. Values of C_E are given in the literature. C_E is a function of electron energy and therefore of the depth in the medium.

In the use of ion chambers, various precautions must be taken.

Saturation. In measuring the charge Q in the ion chamber, not all the ions will be collected due to recombination. This may become a critical problem for electron beams where the dose rates are high and the beams are pulsed. This effect has been analyzed by Boag and experimentally investigated by several groups (Boag, 1966; Ellis and Read, 1969; Greening, 1964). The Hospital Physicists' Association (1971) suggests that the following test can be used to determine if sufficient voltage has been applied to the ion chamber. Take a reading, R_1, with a given monitored dose, using a polarizing voltage V, then repeat the measurement for the same monitor units with only half the polarizing voltage to yield reading R_2. If the percentage change in the reading given by

$$X = \frac{100\ (R_1 - R_2)}{R_1} \quad (11)$$

is less than 5%, then the efficiency at the full polarizing voltage V is $(100-X)\%$. If X is greater than 5%, the polarizing voltage must be increased.

Polarity Effects. The polarity of the polarizing voltage can affect the measured ionization for a number of reasons, including incident electrons upon the collector, variation of the active volume of the chamber with polarity, etc. It is, therefore, necessary to measure the ion current with both negative and positive polarizing voltages and to use the mean of the readings.

Stem and Cable Effects. Irradiation of the stem and cable may cause an unwanted current to be collected. This must be checked, and the necessary precautions (either shielding the cable or placing the stem and cable out of the beam) should be taken.

Perturbation. When the ionization chamber is introduced into a medium, the medium is displaced by a small cavity of air. Electrons scattered into the cavity do not scatter out as easily, and the ionization current is higher than it should be. Harder (1968) has fully discussed this effect for variously shaped cavities and describes a perturbation factor, P_{wg} which, multiplied by the ion chamber reading, corrects for this effect:

$$P_{wg} = \frac{1}{1 + Jgp/Jg} \qquad (12)$$

where Jgp is the additional ion current due to the perturbation effect and Jg is the undisturbed ion current. For thimble ion chambers of radius r perpendicular to the electron beam

$$\frac{Jgp}{Jg} = \frac{2}{5} \frac{br^{1/2}}{\pi} \qquad (13)$$

where for a water medium

$$b = \frac{1.096 \, (E_z + 0.511)}{E_z \, (E_z + 1.022)} \qquad (14)$$

E_z is the energy of electrons at depth Z in the water medium and is given by equation 2.

Table 1 lists the factor P_{wg} for various thimble chambers as a function of E_z. At the higher energies and for small chambers, this effect is small, and P_{wg} is often taken as

TABLE 1. PERTURBATION FACTOR P_{wg}

Perturbation Correction Factor P_{wg} for a thimble ion chamber of inner radius r in water, for different mean electron-beam energy E_z (21, 18).

E_z (MeV)	r=0.25 cm	r=0.5 cm	r=0.8 cm
1	0.947		
2	0.970	0.958	
3	0.978	0.970	0.902
4	0.984	0.977	0.972
5	0.986	0.981	0.977
10	0.992	0.989	0.987
20	0.997	0.995	0.993

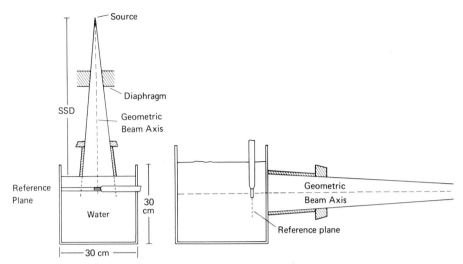

Figure 6. Examples of measuring geometries. The beam may be either vertical or horizontal. Because of the displacement effect, the reference plane will be 0.75 the chamber radius in front of the center of the chamber.

unity. However, towards the end of the electron beam range, when the energy of the electrons approaches zero, this factor can be appreciable.

Displacement Correction. The above corrections are made on the ion chamber reading. This displacement correction is made on the position of the chamber. The point of measurement will be displaced from the center of the chamber towards the source of the electron beam. This has been investigated both experimentally and theoretically, and for all practical situations with cylindrical or spherical ion chambers of radius r, the point of measurement is taken as $(0.75) r$ in front of the chamber center (Dutreix and Dutreix, 1966; Hettinger et al., 1967; Skaggs, 1949).

ABSORBED DOSE CALIBRATION

For the initial calibration the use of a suitable ion chamber in a water phantom is recommended. The ion chamber should be small (up to about 8 mm in diameter and several centimeters in length) and connected to a high-quality, low-current measuring instrument, such as the Townsend balance system, vibrating reed electrometer, or operational amplifier. It should have a ^{60}Co calibration factor (N) traceable to a national standardizing laboratory. All precautions outlined in the above paragraph must be taken into account. In particular, complete saturation must be reached.

The calibration set-up is shown in Figure 6. If the beam is directed horizontally, then the thickness of the entrance phantom wall should be less than 0.5 cm. The ion chamber should be protected during the measurements, either with a waterproof sheath or in a polymethylmethacrylate tube with walls of 1 mm or less. It must be possible, also, to place the chamber at the required depths within the phantom. A standard field size of 10 cm X 10 cm at the surface is suggested for the calibration procedure. The water phantom should be large enough so that none of the electrons can leave it. The International Commission on Radiation Units (ICRU) has recommended a tank 30 cm X 30 cm X 20 cm. If small field sizes and lower energies are to be used, then the dimensions of the tank can be appropriately reduced.

The depth at which the ion chamber is placed is dependent upon the energy of the electron beams. The different protocols have suggested various depths for different energy ranges, but due to the difference in the shape of the depth-dose curves, no standard depth can be listed. The ICRU handbook (Report No. 21, 1972) notes that the depth of maximum dose is not a satisfactory choice because of the presence of scattered electrons originating from the machine and its accessories; it recommends that "the reference depth be sufficiently great so that the contribution of scattered electrons is not serious, but close enough to the maximum so that there is a slow variation of absorbed dose with depth," i.e., the calibration depth should be just beyond the dose maximum. The depth-dose curves should, therefore, be measured for each machine, and the depth of calibration chosen accordingly. For example, the 7.0 and 9.0 MeV beams from a betatron and 7.0 and 10.0 MeV from a linac were measured, with the results shown in Figures 7 and 8. On the betatron the surface levels are 91 and 94% with d_{max} at 10.0 mm and 12.5 mm for 7.0 and 9.0 MeV, respectively; on the linac the surface levels are 79.0 and 88.0% with d_{max} at 14.5 mm and 24.5 mm for 7.0 and 10.0 MeV, respectively.

It is advisable, if possible, to check the absorbed dose calibration with an independent system, and the ferrous sulfate dosimeter can be used for this purpose. The performance of the $FeSO_4$ dosimeter should first be checked in a calibrated high-energy beam (e.g., ^{60}Co γ-rays).

The frequency of the routine dose calibration will depend on the stability of the machine. It is recommended that the machine be calibrated once a week at all energies, but if it is found that variation of more than 3% in the output occurs between calibrations, then they should be done more frequently. Routine calibration can be done in the polystyrene phantom (Fig. 9) known as the SCRAD phantom (AAPM, 1966). The relationship between the calibration factors in this phantom to the dose measured in water should be determined during the initial measurements.

DEPTH-DOSE MEASUREMENT

Central-axis, depth-dose curves have been measured with a variety of dosimeters, including ion chamber, film, solid-state detectors, chemical detectors and biological dosimeters.

By the use of an ionization chamber to scan down the central axis of the beam, a depth versus ionization curve may be obtained. To convert this to a central-axis depth-dose curve, the corrections given in the above section on ion chambers must be applied at each depth. It is assumed that the chamber reading is corrected for saturation, polarity and stem effect. Equation 12 must then be applied for the perturbation effect, and the ion chamber reading multiplied by the appropriate C_E factor to make it proportional to dose. For this purpose, an empirical equation for C_E has been obtained:

$$C_E (E_z) = 0.97 E_z (\exp) -0.048 \qquad (15)$$

and equation 2 can be used to find E_z.

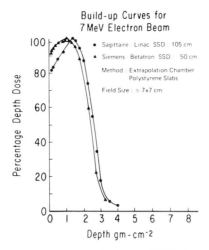

Figure 7. Central-axis depth-dose curves for the 7-MeV electron beams of the Sagittaire linear accelerator and the Siemens betatron, showing the surface dose and build-up region. (From de Almeida and Almond, 1974)

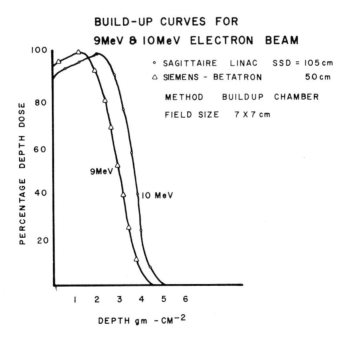

Figure 8. Central-axis depth-dose curves for the 9-MeV electron beam from the Siemens betatron and the 10-MeV electron beam from the Sagittaire linear accelerator, showing the surface dose and build-up region. (From de Almeida and Almond, 1974)

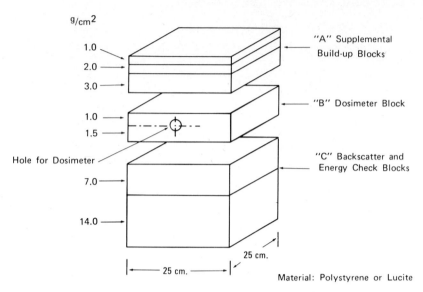

Figure 9. "SCRAD" Calibration Phantom.

Finally, the 0.75 r displacement factor must then be incorporated. The overall effect of these corrections is to make slight corrections to the curves, as shown in Figure 10.

Thermoluminescent dosimeters may also be used to obtain central-axis, depth-dose curves (Almond et al., 1967). These have the advantage that only one exposure need be made. The dose should be chosen, however, so that the thermoluminescent readings obtained are within the linear response range of the dosimeter. The drawback to the TLD method is the length of time required to read the dosimeters.

Film may also be used to determine the central-axis, depth-dose curve. The film can be exposed either perpendicularly or parallel to the beam, but for the former, several films have to be exposed, while for the latter only one film need be used. However, for the parallel film the first few millimeters of data are not reliable, as explained below.

Other detectors, such as diodes, can also be used. Solid-state detectors and film require no corrections to obtain the relative central-axis, depth-dose curves. Figure 11 shows a comparison of the central-axis, depth-dose curve, as determined by a number of systems.

FILM DOSIMETRY

Various studies have shown that film dosimetry is an excellent practical method for use with high-energy electrons. A number of films are available, and the variation of the optical density as a function of absorbed dose (the sensitometric curve) should be measured for each film type. Where possible, films which have a linear relationship between density and dose should be used (Figure 12). However, some authors have used non-linear films and have made the necessary corrections relating density to dose (Van Patten et al., 1974). The best results have been obtained with hand-processed film using the same procedure every time, although automatic processed film can be used in some instances.

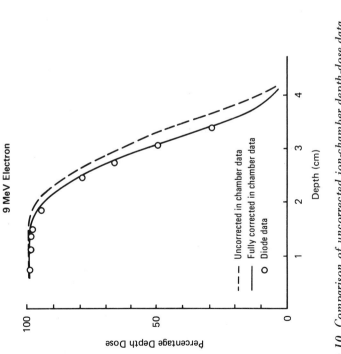

Figure 10. Comparison of uncorrected ion-chamber depth-dose data and fully corrected, depth-dose data. This shift, due to the displacement effect, can be seen. The diode data agree with the corrected ion chamber data.

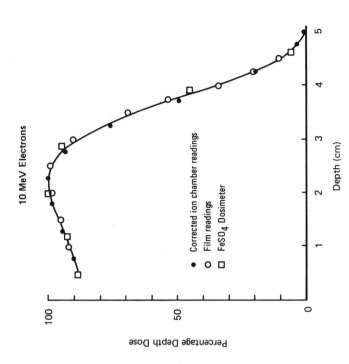

Figure 11. Comparison of central-axis, depth-dose data measured with an ion chamber, film and $FeSO_4$ dosimeter for 10 MeV electrons.

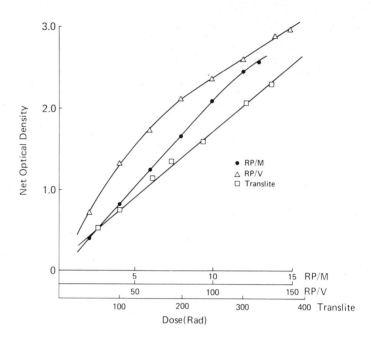

Figure 12. Sensitometric curves for three different types of film. The RP/M and RP/V are rapid processing (automatic processing) film. The translite is hand-developed. The net optical density (density minus background) is plotted versus the absorbed dose.

Dutreix and Dutreix (1969) have listed the accuracies that can be expected with the use of film (Table 2). It should be noted that while film of the same batch can be successfully used, even though processed at different times, variation between batches of the same film can be large enough that the sensitometer curve for each batch should be obtained.

Like other solid-state detectors, the ratio of the collision stopping power in the emulsion and in muscle varies slowly with energy, for energies between 0.3 and 20 MeV. Depth-dose curves obtained with film show excellent agreement with other solid-state detectors and with ion chambers, when the latter measurements are fully corrected. However, for energies above 20 MeV, various groups have observed that film will underestimate the dose near the surface of entry.

Since the electron-beam dose distribution can have areas of very steep gradients, the high spatial resolution obtainable with film makes this method very attractive. However, certain precautions must be taken in using the film. In order to obtain the isodose curve, the film must be exposed parallel to the beam, with the edge of the film even with the surface of the phantom to avoid the difficulties presented in Figure 13 and to allow meaningful distance measurements to be made.

To obtain the isodose curve, the film is exposed in a polystyrene phantom. The special requirements of the phantom are that it must be possible to compress the film pack between two accurately flat surfaces. A diagram of such a phantom is shown in Figure 14 (Feldman et al., 1974). Polystyrene is chosen as the phantom material since

TABLE 2. RELATIVE MEASUREMENTS BETWEEN SERIES OF FILMS PROCESSED SEPARATELY.*

Great care is required in reproducing the processing conditions which are restricted to films of the same batch. The accuracy that can be expected is summarized as follows:

On the same film	2%
On films processed simultaneously	3%
On films processed separately (identical processing conditions)	5%
On films of different batches	5%

*Taken from Dutreix and Dutreix, 1969.

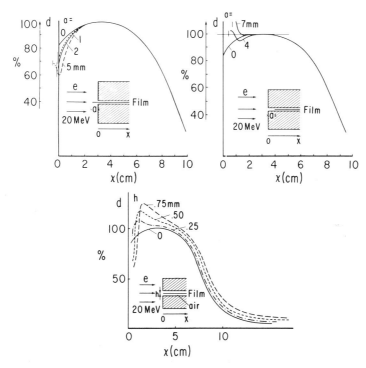

Figure 13. Influence on the depth-density curves due to misalignment of the film in the phantom. Three effects can be observed: (i) if the film extends beyond the phantom, or (ii) if it is recessed within the phantom, and (iii) if there are any air gaps between the film and the phantom. (Redrawn from Dutreix and Dutreix, 1969.)

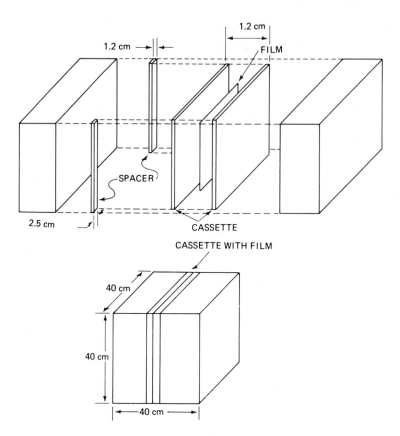

Figure 14. Schematic drawing of a film phantom made of opaque polystyrene. Not shown is the clamping device that holds the phantom together and excludes air trapped adjacent to the film. (From Feldman et al., 1974.)

its electron density per unit volume is very close to that of water and tissue and its specific gravity is also close to unity. The phantom shown is of high impact (opaque) polystyrene with a density of 1.04. Due to this density the resulting isodose curves are adjusted to match the unity-density, central-axis, depth-dose curve, but no other correction is made. Since the phantom is opaque, light generated by the Cerenkov process does not affect the film. The film can be either loaded directly into cassettes made of the phantom material or placed in the phantom in ready-packed containers, if the film is available in this form. The film packs should be punctured in several corners with a sharp point to prevent air from being trapped. If the film wrapping extends a short distance beyond the phantom, the excess should be folded to one side and taped down tightly. The radiation exposure should be such that the maximum density is close to the maximum density on the linear portion of the dose-density curves. After the film is processed, it is analyzed with a densitometer using a light spot of about 1 mm diameter. The process is more rapid if an automatic density-plotter is used.

ISODOSE CURVE MEASUREMENTS

As with the central-axis, depth-dose curves, there are several methods for obtaining electron-beam isodose curves. However, for isodose curves considerably more data have to be obtained, and this should be kept in mind when selecting the method to be used. For example, thermoluminescent dosimeters may be placed in a matrix system in the plane of interest and a single exposure made. However, on a 1-cm grid for a 10 cm X 10 cm field for a 15 MeV electron beam, the matrix would have to consist of at least 84 dosimeters, all with equal or known responses. Diodes or ion chambers can be used with automatic isodose plotters. Use of this technique requires two detectors (or the monitor plus one dectector) since the instantaneous output of electron-beam machines can vary. This means that integrated measurements must be made at each point, and the time required to obtain a single curve may be quite long.

Because of these considerations, film dosimetry has considerable advantages over other systems, as well as allowing high spatial resolution to be obtained. Since the isodose curves can have areas of very steep dose gradients, the high spatial resolution that can be obtained with film makes this method very attractive. Certain precautions must be observed in using the film. For the isodose curve to be obtained, the film must be exposed parallel to the beam with the edge of the film even with the surface of the phantom to avoid the difficulties discussed above.

After the film is processed, it is analyzed with a densitometer, using a light spot of about 1-mm diameter. The densities are read out on a matrix, and the isodose curve drawn from these data. The process is more rapid if an automatic density plotter is used. The only correction that is made is the adjustment of the central axis to agree with the final, central-axis, depth-dose curves obtained with an ion chamber or a diode in water. Typical isodose curves are shown in Figure 15.

BEAM CONTROL

In this section monitoring of the beam will be discussed first, since the incident fluence of electrons can be directly related to absorbed dose in the medium, and monitoring the fluence is necessary to ensure the correct patient dose.

Secondly, factors affecting the dose distribution will be discussed; these are flattening foils or scanning magnets to spread the beam and beam collimation.

MONITORS

As mentioned above, it is necessary to monitor the electron fluence leaving the machine in order to control the dose to the patient. A number of instruments have been used, including transmission ionization chambers, secondary electron emission monitors, electromagnetic induction systems, and sampling methods.

The most common type of monitor is a transmission ion chamber, which generally consists of a thin parallel-plate ion chamber through which the total beam passes. Several foils are used so that there are two transmission chambers, with one of the chambers divided into halves or quarters. By comparing the signals from both chambers, a malfunc-

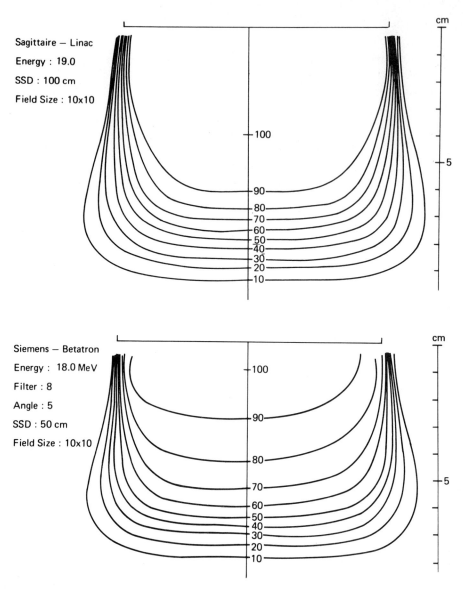

Figure 15. Examples of isodose curves measured with film. (From de Almeida and Almond, 1974.)

tion in one chamber can be detected, and by comparing the signal from the two halves (or the four quarters), asymmetry in the beam can be detected. From this viewpoint the transmission chambers are most useful, but they may interfere with use of the normal light localizers. Some machines have transmission chambers that are removed when the light localizer is used, but this requires precise relocation of the chamber for each treatment. Transmission chambers also act as extra scattering and absorbing material, causing a reduction in the energy and introducing low-energy electrons into the beam. The monitors are also affected by ionic recombination in the chamber. When high-pulsed currents are used, there may be a lack of saturation in the chamber. For this reason, high collecting voltages should be used, and the separation of the chamber foils should be as small as possible. Even so, with present-day accelerators the collection efficiency may still be less than 100%. This means that the monitor should be calibrated at the dose rates used for therapy, since the collection efficiency will change with dose rate. This type of measurement should also be made to check the linearity of the monitor and to be sure that it is zeroed correctly. When these measurements are made, leakage currents in the monitor chamber should also be checked. Besides monitoring total dose, the transmission chambers also monitor instantaneous dose rates. Due to the possibility of very high dose rates with electrons, the monitoring system should be equipped so that the machine is turned off if the dose rate exceeds a given value.

Although transmission ion chambers are the most common type of monitor, ion chambers have also been used as sampling monitors to look at a fraction of the primary beam or scattered beam. This method depends on the fluence at the chamber bearing a constant ratio to that at the point where the radiation is being used. Introduction into the beam of scattering foils or different collimators, etc., may change this ratio. In general, the total beam monitor is the safest.

Most chambers are sealed so that the amount of air in them is constant. If they are not, then the calibration factor will change with temperature and pressure, and the calibration can vary as these parameters change. It is, therefore, advisable to check that the monitor chamber is sealed on a regular basis.

FIELD FLATNESS

Scattering Foils. So that a uniform dose distribution across the treatment field can be obtained, scattering foils are placed in the beam. Figure 16 shows the relationship between the initial electron beam, which generally has a cross-sectional diameter of a few millimeters, the scattering foils, and the surface to be irradiated. The electrons are scattered through an angle from the central axis, and it is required that the flux off-axis out to C be equal to the flux on-axis at A. At the edge of the field the flux will be down to some fraction of the central flux, usually 0.95. The probability that the electron will be scattered through the angle θ is proportional to $exp-(\theta/\bar{\theta}_o)^2$. $\bar{\theta}_o$ is the root mean scattering angle. With the field edge at C and the design goal that the flux at C be 0.95 of that at A, the ratio of the flux at C to A (taking into account inverse square) is given by:

$$\frac{exp-(\theta/\bar{\theta}_o)^2}{1+\tan^2\theta} = 0.95 \qquad (16)$$

This equation can be solved for $\bar{\theta}_o$, and for a given element the thickness X of the foil can be calculated. However, the energy losses in the foil, in particular, the radiation losses must be kept down to a minimum. For a foil of thickness X, radiation length lr, for an initial electron energy of E_o, the radiation loss is approximately:

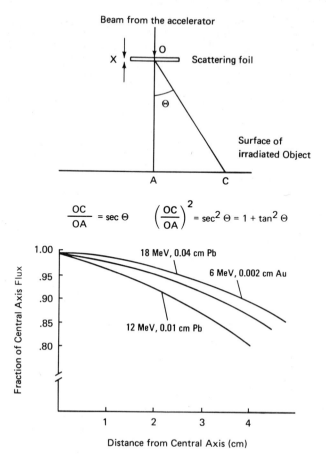

Figure 16. Schematic drawing for scattering foils and results of calculations for various electron energies and typical foils. For the 18-MeV beam the collision energy loss in the foil was 0.58 MeV and the bremsstrahlung loss was 1.4 MeV. For 12 MeV the losses were 0.14 MeV and 0.23 MeV, respectively, and for 6 MeV the losses were 0.046 MeV and 0.04 MeV.

$$E_{rad} = \left(\frac{X}{lr}\right) E_o \qquad (17)$$

Figure 16 shows the results of such calculations for energies and foils of typical thickness and composition used in the therapy machines.

It should be recognized that besides introducing x-rays into the field, the foils also introduce secondary electrons of lower energy, which affect the build-up region, central-axis, depth-dose shape, and energy of the beam.

Scanning Magnets. To overcome the difficulties of using scattering foils, it is possible to scan the initial electron beam to cover larger areas. The first approach to this was with a thin beam scanned in a set pattern over the area to be treated (Lanzl, 1968). However,

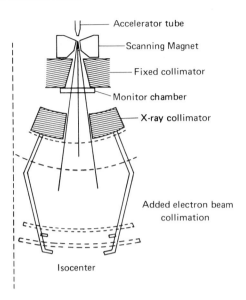

Figure 17. Collimator geometry.

the latest approach has been to use a broad beam and scan in a random fashion. A quadrapole scanning magnet is used, with one pole pair alternating its magnetic field at a frequency different from the other pair. In this way, a very uniform distribution is reached in a short time. The scanning magnets do not affect the beam energy or introduce low-energy electrons into the beam.

COLLIMATION

The aim of a good collimation is to provide a useful clinical machine. This requires that a variety of field sizes be available and that they are easily obtained. The size of the collimators should be such that they can be used in various clinical situations. Small field sizes are often used in the head and neck region, so that the collimator should not be too bulky and should not interfere with anatomical structures. Larger field sizes (up to 25 cm X 25 cm) are generally used for chest-wall treatments, where the size of the collimator is less critical.

For electron-beam therapy, variable-field or fixed-field collimators are used. In general, all collimators follow these principles—a primary collimation close to the source, which defines the field size, and a secondary collimation close to the patient (Fig. 17). Some collimators are open on the side, whereas others are completely enclosed. Figure 18 shows examples of both types of collimators. Various studies of the materials to be used for collimation have shown that materials with medium and high atomic numbers are the

Figure 18. (a) Photograph of the Sagittaire electron-beam collimator illustrated in Figure 17. It can be seen that the collimators are open on the side. The lower bars define the field close to the patient, while the upper bars prevent scattered electrons from reaching the patient.

best. Dahler (1964) and others (Svensson and Hettinger, 1967) have shown that high Z material close to the patient is necessary to sustain a uniform distribution over the field for low- to medium-energy electrons. For high-energy electrons above 20 MeV, E. Briot et al. (1973) have found that the collimation close to the patient is not necessary.

Figure 18. (b) Siemens treatment cone. This applicator is made from brass and is completely enclosed. The cone is brought down to the patient surface. The knurled knob above and to the right of the cone is the device for changing the angle of the cone relative to the beam direction.

The overall effect of scattering foils and collimation should be viewed from two aspects: effects upon central-axis, depth-dose curves, and effects upon the isodose distribution.

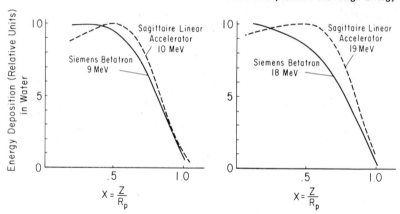

Figure 19. Comparison of central-axis, depth-dose curves measured in water for an accelerator with open collimator and scanning magnet (Sagittaire) and closed collimator and scattering foils (Siemens). The relative depth-dose is plotted versus normalized depth $X = \dfrac{Z}{Rp}$ to remove the slight difference in the energy.

RESULTS

Introduction of lower-energy scattered electrons into the beam from the scattering foils, transmission chamber, and collimators, etc., will have the effect of increasing the surface dose and moving the maximum towards the surface. The effects of scattering foils and a closed collimator can be seen in Figure 19, which compares data from a machine using such a collimator and scattering foils to data taken with scanning-magnet and open collimators. The maximum dose is moved much closer to the surface, the surface dose is increased, and the fall-off with depth is less sharp.

The effects on the central-axis, depth-dose curves will be reflected in the isodose curves, and if the collimation is not optimized, the distribution across the beam will be affected. This is illustrated in Figure 20. Too little collimation results in non-uniformity across the surface, whereas too much collimation produces hot spots at the edge of the beam.

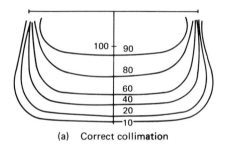
(a) Correct collimation

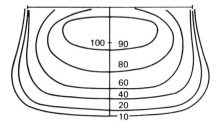
(b) Too little collimation

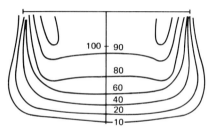
(c) Too much collimation

Figure 20. The effects of collimation upon the isodose curves. (a) With correct alignment of the collimator and the beam, a uniform distribution is obtained. (b) With too little collimation, i.e., the lower edges of the collimators are farther apart than in (a), the isodose lines are pulled in from the field edge, resulting in more dose to the center of the field than towards the edge. (c) The opposite effect—too much collimation. With the lower edges of the collimator being closer together than in (a), too many electrons are scattered from the collimator into the field, adding significantly to the dose around the edge. This results in "hot spots" as shown.

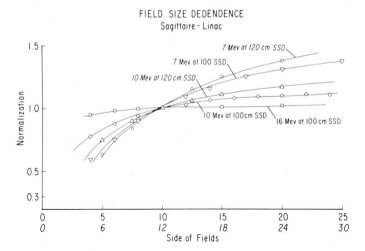

Figure 21. Field-size dependence curves for machine output. The curves are normalized to the 10 cm X 10 cm field size: on the abscissa (sides of the fields in centimeters) the upper numbers are for the 100 cm S.S.D. and the lower for the 120 cm S.S.D. With increasing energy the field-size dependence decreases. (From de Almeida, Master's Thesis, 1973.)

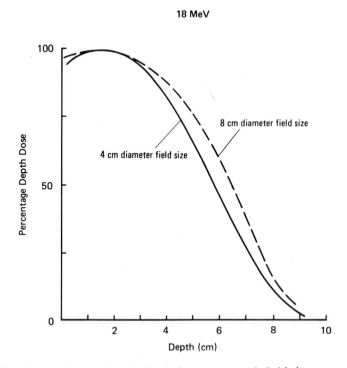

Figure 22. The changes in central-axis, depth-dose curves with field size.

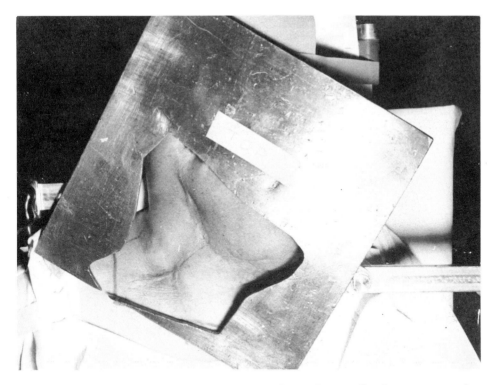

Figure 23. Typical shaped field used in electron-beam therapy. Lead cut-outs are often used to shape the field in the neck region.

FIELD-SIZE DEPENDENCE

The output and the central-axis, depth-dose curves are dependent on field size. As noted above, the dose on the central axis is made up from energy deposited by the primary electrons and the scattered electrons. The dose will, therefore, increase with field size, as long as the distance between the point of measurement and the edge of the field is shorter than the range of the secondary electrons. When this distance is reached, the central-axis, depth-dose curve and the output become constant with field size. Because the energy and direction of the scattered electrons are very much dependent on the energy of the primary electrons, the field-size dependence is also a function of energy. Figures 21 and 22 show these effects.

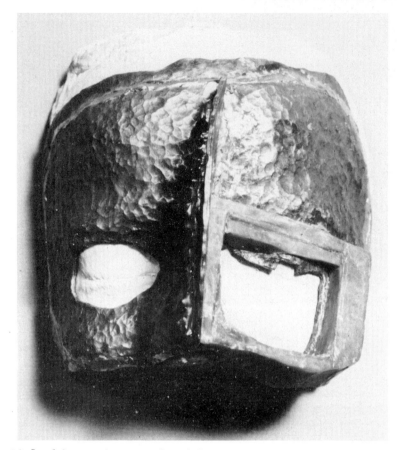

Figure 24. Lead face masks are used to define the beam around eyes and nose to protect the eyes.

SPECIAL CALIBRATIONS

There will be clinical situations that are not covered by the routine calibrations. These arise when irregular fields are used or when special blocking devices are made, often using lead to define the field. Figures 23 and 24 show a typical blocked field and a face mask made of lead to protect the face when an eyelid is being irradiated. There are several considerations when these kinds of procedures are followed.

THICKNESS OF LEAD REQUIRED TO REDUCE THE BEAM

The thickness of the lead shielding is usually chosen to reduce the dose by 95% to 98% relative to the open beam. Table 3 shows thicknesses of lead for different electron energies and field sizes. For high-energy electrons, additional lead beyond a few millimeters will not decrease the dose to 98% or even 95% levels, due to the bremsstrahlung produced.

SAGITTAIRE

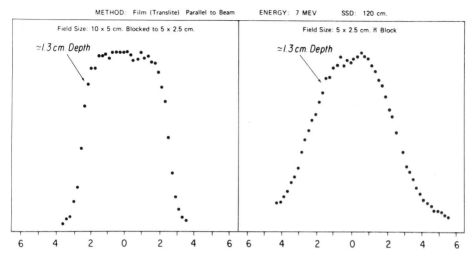

Figure 25. Improvement of beam flatness of the Sagittaire linear accelerator at a depth of 1.3 cm when fields are blocked. Factors of 7 MeV and 120 cm SSD were used with Translite film parallel to the beam. Plotted is the net optical density (proportional to dose) versus the distance across the field in centimeters. (From de Almeida and Almond, 1974.)

TABLE 3. THICKNESS OF LEAD REQUIRED TO REDUCE THE BEAM

Field Size (cm)	Attenuation	Lead Thickness (mm)				
		7 MeV	9 MeV	11 MeV	15 MeV	18 MeV
6 X 6	98%	2.6	4.0	5.0	*	*
10 X 10	98%	2.8	4.0	*	*	*
6 X 6	95%	2.3	3.3	3.8	5.4	7.6
10 X 10	95%	2.4	3.3	4.3	6.0	*

*Not obtainable because of Bremsstrahlung

POSITION OF THE LEAD

When a field is blocked with lead so that the desired field shape is obtained, it is important that the lead be placed near the surface of the patient to obtain a reasonable flatness of the beam across the field. The effect of using lead to block the field is shown in Figure 25. (From de Almeida and Almond, 1974.)

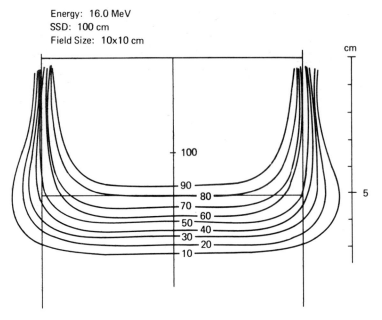

Figure 26. Illustration of the constriction of the 80% isodose line at the depth of the 80% dose. The field size at the patient surface is projected down to the 80% depth, and it can be seen that this area must be reduced by 1 cm on all sides to give an adequate dose distribution to the treatment area.

DOSE CALIBRATIONS

In some cases it may be necessary to carry out special calibrations. If the field is large and the blocking is minimal, then the normal calibration for the field size set by the collimator can be used. The field-size dependence curve should be used to determine if the output is constant over the areas used. If it is not, or if the blocking is over more than 25% of the field area, a special calibration should be carried out. For most cases, a direct calibration can be made with use of the SCRAD phantom with the field and blocking set-up to reproduce the clinical situation. If the field is very small, or the body surfaces are very irregular, such as is the case with face masks, then a wax mold should be used. Small thermoluminescent dosimeters can then be placed at the build-up depth in the wax, and the clinical set-up reproduced for the calibration.

IN-VIVO DOSIMETRY

In vivo measurements in electron-beam radiation therapy have proved very useful. For practical reasons, thermoluminescent dosimeters have become the dosimeter of choice and are now used extensively. Dose distribution in electron-beam irradiation to the head and neck region is of considerable interest due to the presence of irregular air-filled cavities.

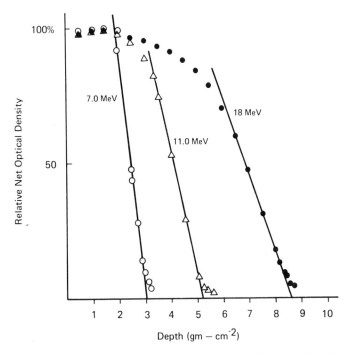

Figure 27. Angled incident beam, showing isodose curves following the phantom surface.

Figure 26 shows a patient who was treated with a 15 MeV electron beam through a 6 cm X 5 cm cone placed appositional to the bridge of the nose to a daily given dose of 250 rads. A small polyethylene tube containing 9 LiF dosimeters separated by small lead shot for identification on AP and lateral films was inserted along the floor of the nasal cavity and brought out through the mouth. The portion within the oral cavity was held in position along the undersurface of the palate by means of a dental stent. Standard dosimeters were also exposed in the calibration phantom at the position of maximum buildup to a dose of 200 rads. During a subsequent treatment, the procedure was repeated; this time the capsules were placed in the upper portion of the nasal cavity and the roof of the nasopharynx. The results of the two measurements are shown with readings as percentages of the given dose. Due to the high dose in the center of the field, the treatment was suitably adjusted, and due to the irregular distribution, a subnasal field was also employed to smooth out the distribution.

It should be pointed out that in using thermoluminescent dosimeters, the standard dosimeter should be exposed to the same energy beam and to approximately the same dose as those in the *in vivo* measurement. Further, their shape must be the same as those used *in vivo* to avoid energy, configuration, and dose effects.

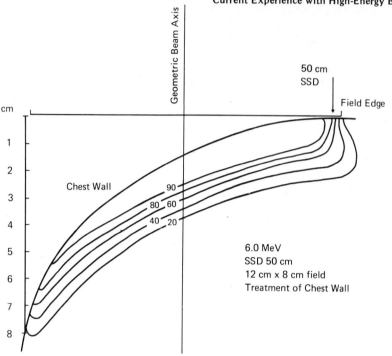

Figure 28. Typical chest-wall field. The isodose curves follow the contour of the chest wall. Corrections have been made for dose fall-off on the lateral edge of the field due to inverse-square effects.

TREATMENT-PLANNING PROCEDURES—HOMOGENEOUS TISSUES

For homogeneous tissues and a relatively flat surface, treatment-planning is straightforward, and the dose distribution can be found with the appropriate isodose or central-axis, depth-dose curve. The energy of the electron beam is often chosen so that the tumor lies entirely within the 80% isodose curve. However, it should be pointed out that the area over which the 80% isodose curve extends, at the 80% depth, is considerably less than the field area on the surface. This is shown in Figure 26 and is called the "constriction" of the isodose curve at depth. The margins of the field at the 80% depth range from 70% to 50%. Therefore, to cover an area by the 80% curve, a proportionately larger field at the surface must be used.

When the distance of the patient from the source is increased, the change in dose will follow the inverse-square law, if the source position is taken as the virtual source. There will be a change in relative depth-dose curve consistent with the geometric divergence of the beam.

When the entry surface is not perpendicular to the beam axis, it is found that the isodose curves are parallel to the surface, and the absorbed dose distribution along the central axis is the same as for a normally incident beam (Fig. 27). The same applies for smooth surfaces with large radii or curvature, which make the electron beam very suitable for treating the chest-wall, as illustrated in Figure 28.

EXPERIENCE WITH ELECTRON IRRADIATION OF BREAST CANCER

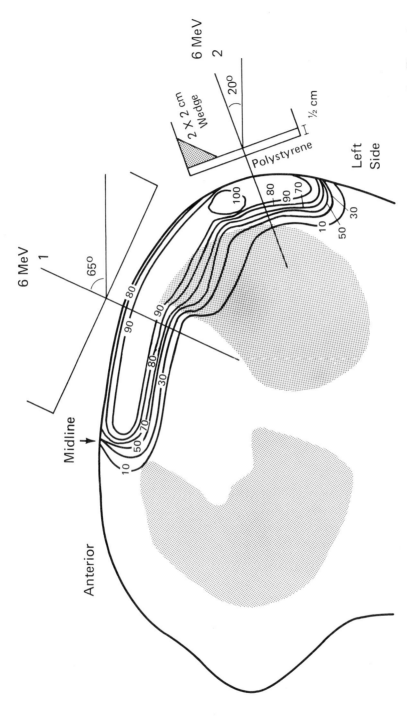

Figure 29. Case illustrating the use of polystyrene wedge to reduce hot spots for two adjoining fields coming in at slight angles. The polystyrene slab is used to reduce the incident 6 MeV electrons by approximately 1 MeV, so that the lung behind the chest wall will be spared. (Redrawn from Chu, 1965).

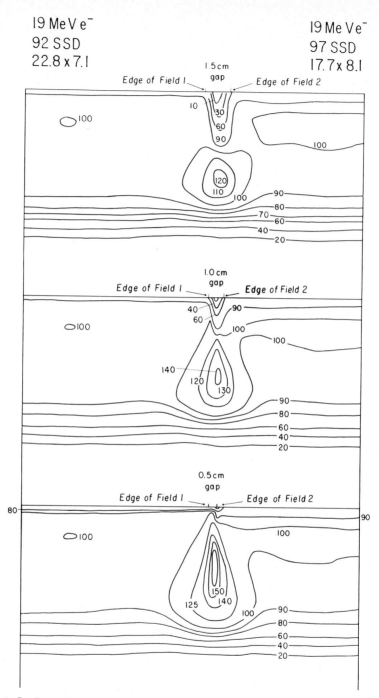

Figure 30. Isodose distributions for adjoining fields with the same electron-beam energy with different gap-widths between the fields.

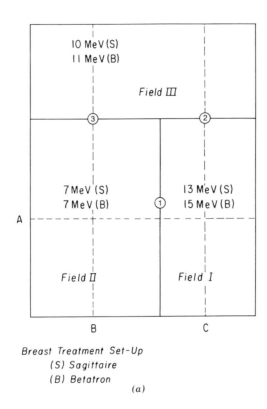

Figure 31. (a) Schematic drawing of the electron-beam field used for typical chest wall (II), supraclavicular (III), and internal mammary (I) treatment. (Reprinted from de Almeida, Master's Thesis, 1973).

Sharp irregular surfaces, on the other hand, cause complex dose-distribution patterns, with the electrons being scattered out, by steep projections, and in, by steep depressions. For these situations it may be possible to smooth out the irregularity with bolus. Absorbers of different thickness can be used to reduce the penetration of the electrons in part of the beam, or to reduce the energy of the total beam. Figure 29 illustrates this application, where the absorber reduces the energy of the beam so that the lung will be spared (Chu, 1965).

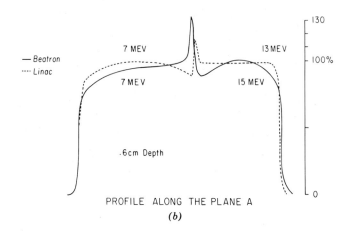

Figure 31. (b) The dose distribution at a depth of 0.6 cm through the plane A. (Reprinted from de Almeida, Master's Thesis, 1973).

It is quite common when using electrons in radiation therapy to adjoin fields. When this is done, care has to be taken that undue hot or cold spots do not result in the distribution. If the fields overlap too much, then the bulging in the isodose curves results in hot spots in the distribution, but if the fields are separated, then a cold spot will occur toward the surface. The combined isodose distributions for adjoining fields of the same energy are shown in Figure 30 for different field separations. It is also common to use different energies for each field, and typical distributions at depth are shown in Figure 31. Even when the fields abut each other, hot and cold spots will appear. To overcome this, wedges, usually made from polystyrene, can be used. Great care must be taken with wedges, since the scattering of electrons from the wedges can cause significant changes in the dosimetry. If the wedge is too far removed from the surface, the multiple scattering of electrons tends to reduce the effect of the wedge. When the wedge is in close contact with the patient surface, it acts as bolus.

ELECTRON-BEAM PROTOCOLS

In the past few years, four protocols concerning electron beam dosimetry have been written:

>American Association of Physicists in Medicine. 1966. Protocol for the dosimetry of high-energy electrons. *Phys Med Biol* 77:505.
>
>Hospital Physicists' Association. 1971. A practical guide to electron dosimetry (5–35 MeV).
>
>International Commission on Radiation Units and Measurements. 1972. Report 21, Radiation Dosimetry: Electrons with Initial Energies between 1 and 50 MeV.
>
>Nordic Association of Clinical Physics. 1971. Procedures in Radiation Therapy Dosimetry with 5 to 50 MeV Electrons and Roentgen and Gamma Rays with Maximum Photon Energies between 1 and 50 MeV.

These protocols have provided many useful suggestions for standardization of the dose-calibration procedures and measurement and specification of beam parameters.

REFERENCES

Almond, P.R., A. Wright & J.F. Lontz II. 1967. The use of lithium fluoride thermoluminescent dosimeters to measure the dose distribution of a 15 MeV electron beam. *Phys Med Biol* 12:389–394.

Berger, M.J. & S.M. Seltzer. 1964. Tables of Energy Losses and Ranges of Electrons and Positrons. NASA SP-3012. National Aeronautics and Space Administration, Washington, D.C.

Berger, M.J. & S.M. Seltzer. 1966. Additional Stopping Power and Range Tables for Protons, Mesons, and Electrons. NASA SP-3036. National Aeronautics and Space Administration, Washington, D.C.

Berger, M.J. & S.M. Seltzer. 1969. Quality of radiation in a water medium irradiated with high-energy electron beams. p. 127. *In* Book of Abstracts. 12th International Congress of Radiology, Tokyo, Japan.

Boag, J.W. 1966. Ionization chambers. p. 1–72. *In* Attix, F.H. & W.C. Roesch (eds.) Radiation Dosimetry. 2nd edition. Academic Press, Inc., New York.

Briot, E., A. Dutreix, J. Dutreix & A. Penet. 1973. Etude experimentale de la collimation des faisceaux d'electrons par un diaphragme de plomb reglable. *J Radiol Electrol Med Nucl* 54:39–46.

Chu, F.C.H. 1964. Experience with electron irradiation of breast cancer. p. 343–348. *In* Zuppinger, A. & G. Poretti (eds.) Proc. International Symposium on High Energy Electrons, Montreux. Springer-Verlag, Berlin.

Dahler, A. 1964. Effect of collimator-shape on electron depth dose curve. p. 98. *In* Proc. International Symposium on High Energy Electrons, Montreux. Springer-Verlag, Berlin.

de Almeida, C.E. & P.R. Almond. 1974. Comparison of electron beams from the Siemens Betatron and the Sagittaire Linear Accelerator. *Radiology* 111:439–445.

de Almeida, C.E. & P.R. Almond. 1974. Energy calibration of high energy electrons using a Cerenkov detector and a comparison with different methods. *Phys Med Biol* 19: 476–483.

Dutreix, J. & A. Dutreix. 1966. Etude comparee d'une serie de chambres d'ionisation dans des faisceaux d'electrons de 20 et 10 MeV. *Biophysik* 3:249.

Dutreix, J. & A. Dutreix. 1969. Film dosimetry of high-energy electrons. *Ann NY Acad Sci* 161:33–43.

Ellis, R.E. & L.R. Read. 1969. Recombination in ionization chambers irradiated with pulsed electron beams; I: Plane parallel plate chamber. *Phys Med Biol* 14:293–304.

Feldman, A., C.E. deAlmeida & P.R. Almond. 1974. Measurements of electron-beam energy with rapid-processed film. *Med Phys* 1:74–76.

Greening, J.R. 1964. Saturation characteristics of parallel-plate ionization chambers. *Phys Med Biol* 9:143–154.

Harder, D. 1964. Energiespektren schneller Elektronen in verschiedenen Tiefen. p. 26. *In* Zuppinger, A. & G. Poretti (eds.) Proc. International Symposium on High Energy Electrons, Montreux. Springer-Verlag, Berlin.

Harder, D. 1968. Einfluss der Vielfachstrelung von Elektronen auf die Ionization in gasgefullten Hohgraumen. *Biophysik* 5:157.

Hettinger, G., C. Pettersson & H. Svensson. 1967. Displacement effect of thimble chambers exposed to a photon or electron beam from a Betatron. *Acta Radiol TPB* 63: 61–64.

Hospital Physicists' Association (Radiotherapy Physics Committee). 1971. A Practical Guide to Electron Dosimetry, 5–35 MeV. H.P.A. Report Series No. 4, London.

International Commission of Radiation Units and Measurements. 1972. Radiation Dosimetry: Electrons with Initial Energies between 1 and 50 MeV. Report No. 21. ICRU, Washington, D.C.

Lanzl, L.H. 1968. Electron pencil beam scanning and its application in radiation therapy. p. 55–66. *In* Vaeth, J.M. (ed.) Frontiers of Radiation Therapy and Oncology. Vol. II. S. Karger, Basel, New York.

Karzmark, C.J. Physical aspects of whole body superficial therapy with electrons. p. 36–54. *In* Vaeth, J.M. (ed.) Frontiers of Radiation Therapy and Oncology. Vol. II. S. Karger, Basel, New York.

Nordic Association of Clinical Physics. 1972. Procedures in radiation therapy dosimetry with 5 to 50 MeV electrons and roentgen and gamma rays with photon energies between 1 and 50 MeV. *Acta Radiol TPB* 11:603–624.

Skaggs, L.S. 1949. Depth dose of electrons from the Betatron. *Radiology* 53:868–874.

Sub-Committee on Radiation Dosimetry (SCRAD) of the American Association of Physics in Medicine. 1966. Protocol for the dosimetry of high energy electrons. *Phys Med Biol* 11:505–520.

Svensson, H. & G. Hettinger. 1967. Influence of collimating systems on dose distribution from 10 to 35 MeV electron radiation. *Acta Radiol TPB* 6:404.

Van Patten, L., J.A. Purdy & G.D. Oliver. 1974. Advance abstract for AAPM Annual Meeting: Automated film dosimetry. *Med Phys* 1:110.

Clinical Experience
with Electron Beams

Norah duV. Tapley, M.D.,
Professor of Radiotherapy,
The University of Texas System Cancer Center
M.D. Anderson Hospital and Tumor Institute,
Houston, Texas

GENERAL CONSIDERATIONS

The physical qualities of the electron beam make it a unique therapeutic tool to be used selectively, either alone or in combination with the photon beam. No difference in the biological effectiveness of electrons compared with megavoltage photons has been demonstrated in laboratory studies, the RBE being close to one (Sinclair and Kohn, 1964). In treating more than 3,500 patients with the 7- to 19-MeV electron beam at the M.D. Anderson Hospital from 1963 to 1975, patterns of use have evolved for clinical situations not as satisfactorily managed with the photon beam alone.

The characteristics of the electron beam make it of distinct advantage in treating malignant lesions located at a limited depth. A sterilizing dose of irradiation can be delivered to the tumor, while the total volume of tissue included in the high-dose range is sharply limited. Only when the lesion is superficial and a low energy can be used, or when the tumor dose is kept relatively low (under 5,500 rads), is the electron beam used alone. In all other situations, because of the intensity of the skin reaction and the degree of subcutaneous fibrosis, electrons are combined with photons in the range of either 4 MeV or 18-25 MeV.

Treatment aims and therapeutic principles are based upon the basic parameters of radiation therapy that have been established for megavoltage photon irradiation. The alterations in the accepted time-dose relationships are minimal. Treatment techniques are modified only as dictated by the special characteristics of the electron beam. The selection of the various energies, the combining of electrons and photons, and the ratio of given doses of each beam depend upon the maximum depth to be treated.

A variation of combination treatment is that of additional or "boost" therapy through reduced portals, in order to increase the total tumor dose to the initial gross disease. This practice is predicated on the hypothesis that microscopic disease can be permanently controlled with doses on the order of 5000 rads in 5 weeks, but gross cancer often requires 7000 rads, too high a dose for a large volume. Tolerance is not exceeded when the electron beam is used through reduced fields.

CLINICALLY IMPORTANT ELECTRON-BEAM CHARACTERISTICS

The characteristics of electron beams of 7 to 25 MeV, which are of particular significance in clinical applications, are:

1. Rapid build-up of dose,
2. Depth of high dose dependent upon electron energy,
3. Sharp fall-off in dose beyond 80% point,
4. Constriction of the isodose curves at depth, and
5. Perturbations of dose distribution because of tissue inhomogeneities.

The method used to flatten the beam, either a scattering foil (betatron) or a quadrupole scanning magnet (Sagittaire 25-MeV accelerator), and the collimation devices of the equipment significantly affect the depth of build-up, the percentage depth-dose, and the constriction of the isodose curves.

This investigation was supported by Public Health Service Research Grants CA-06294, and CA-05654 and Training Grant CA-05099 from the National Cancer Institute.

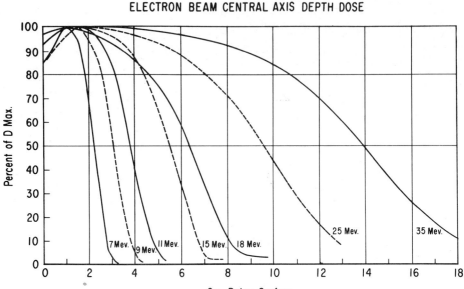

Figure 1. *Central-axis depth-dose curves of 7, 9, 11, 15, 18, 25 and 35 MeV electron beams. The depth-dose falls off sharply, particularly at the lower energies, and the surface dose increases with increasing energy. The curves were derived from available betatron data of the 6- to 18-MeV Siemens betatron (7- to 18-MeV curves) and the 35-MeV Brown-Boveri betatron (25- and 35-MeV curves). The table below gives the depth for the 80, 50, and 10% doses for various electron-beam energies:*

	7 MeV	9 MeV	11 MeV	15 MeV	18 MeV	25 MeV	35 MeV
80%	1.8 cm	2.6 cm	3.2 cm	4.4 cm	4.7 cm	7.3 cm	10.7 cm
50%	2.2 cm	3.1 cm	3.8 cm	5.5 cm	6.4 cm	9.6 cm	13.9 cm
10%	2.7 cm	3.8 cm	4.8 cm	6.7 cm	8.1 cm	12.7 cm	18.1 cm

(Reprinted by permission from Tapley, 1976.)

Within the first millimeter of tissue, the dose approaches 90% of the dose at depth of maximum build-up, which explains the modest skin-sparing of the electron beam. The surface dose increases as the energy increases and approaches 100% for 18 MeV (de Almeida and Almond, 1974). The higher the MeV, the more intense the skin reaction will be because of the higher (more rapid) superficial build-up with the higher energies. There is relatively less skin-sparing with a scattering foil than with scanning magnets, which do not affect the beam energy or introduce low-energy electrons into the beam (Almond, 1975).

A high dose level is maintained to a depth determined by the energy of the beam, with the doses falling off sharply beyond this depth. With the 6- to 18-Mev Siemens betatron, the 80% dose level is reached at 1.8 cm with 7 MeV and at 4.7 cm with 18 MeV (Fig. 1).

Clinical Experience with Electron Beams

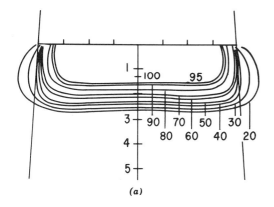

(a)

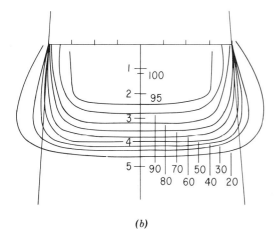

(b)

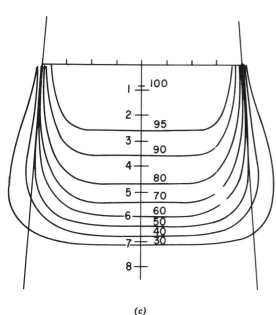

(c)

With the scanning magnet (Sagittaire), the depth of maximum dose build-up is greater, the 80% dose for 7 MeV being at 2.1 cm and for 19 MeV, at 6.7 cm.

There is a significant tapering of the 80% isodose curve (Fig. 2) at energies above 9 MeV. With the betatron, the area included by the 80% curve at depth relative to the field at the surface cannot be increased, even with optimal cone design, and is worse for small fields at the higher energies.

The sensitivity of high-energy electron dose distribution to the presence of tissue heterogeneities (Almond et al., 1976; Boone et al., 1967) makes it essential to consider these effects in treatment planning.

In situ measurements (Boone et al., 1965) have shown a marked increase in percentage depth-dose for both 9- and 18-MeV electron irradiation across lung tissue, when compared with water phantom values for the same depth.

Figure 2. Isodose curves for 7-, 11-, and 18-MeV electron beams obtained with the Siemens betatron. The field size is an 8-cm circle.

(a) 7 MeV. The 80% curve is at 1.8 cm. There is little radiation beyond 2 cm.

(b) 11 MeV. The 80% curve is at 3.2 cm.

(c) 18 MeV. The 80% curve is at 4.7 cm. The decrease in percentage depth-dose is less rapid than at 7 MeV, but little radiation extends beyond 8 cm. Note that the 80% isodose line is not flat at the geometric edge of the field. A generous field must be used. (Reprinted by permission from Tapley, 1976).

The range of the electrons within different materials depends primarily on the electron densities of those materials. The density of compact bone, as in the mandible adjacent to the tonsillar area, is taken to be 1.85 g/cm^3. Considering the ratio of the electron density of bone to that of muscle, 1 cm of bone should produce approximately the same attenuation as 1.65 cm of muscle. With lithium fluoride dosimeters in patients undergoing treatment for lesions in the tonsillar area (Almond et al., 1967), the average percentage decrease in dose, compared with the water phantom value, was approximately 10%. No correction for bone absorption need be applied in the average patient for 11- and 15-MeV irradiation when the sternum (predominantly spongy bone) is interposed in the beam.

NORMAL TISSUE REACTIONS

During the initial experience with electron-beam therapy at the M.D. Anderson Hospital, skin tolerance tables were developed correlating the factors of dose, time, areas, MeV, and anatomical site, including the lateral face, the upper neck, the lower neck, and the chest wall (Tapley and Fletcher, 1965). Table 1 gives the tolerance doses for treatment of small fields on the lateral face and large fields on the neck and the chest wall with electrons of the energies routinely used in those areas. Above these doses, a moist reaction may occur. If the tolerance doses are exceeded, the risk of late changes must be accepted. When treatment is given for lesions where the minimum tumor dose must be deeper than 3-3.5 cm, the photon beam (^{60}Co or 18- to 25-MeV photons) must be added to avoid exceeding the tolerance dose of the electron beam.

TABLE 1. ELECTRON BEAM TOLERANCE DOSES CORRELATED WITH ANATOMICAL LOCATION, FIELD SIZE, AND MEV

Anatomical Location	Field Size	MeV	Dose and Time	Reactions
Lateral face	< 50 cm^2	15–18	6500 rads 6–6½ weeks	Erythema and usually dry desquamation
Neck and chest wall	> 50 cm^2	7–11	5000–5500 rads 4 weeks	Occasional moist desquamation, patchy and superficial No, or minimal, late changes

The acute mucous-membrane reactions seen with the electron beam are similar to those produced by the photon beam for the same doses, but they are sharply localized to the ipsilateral side. Limiting the high dose to one side of the oral cavity insures less discomfort for the patient, improved nutrition during the course of therapy, and decreased interference with salivary gland function. When the electron beam is combined with the 18- to 25-MeV photon beam in the treatment of more deeply located lesions, the mucosal reac-

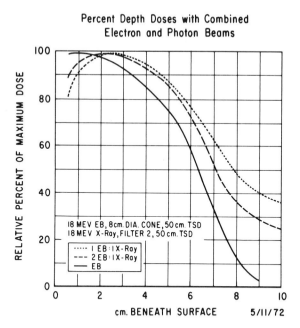

Figure 3. Combined 18-MeV electron- and photon-beam depth-dose curves. With a 1.1 ratio of electron and photon given doses, the calculated 80% dose level is located more than 1 cm deeper than with 18-MeV electrons alone. (Reprinted by permission from Fletcher, 1973)

tion on the side of the entering beam will be patchy or confluent exudate, whereas the opposite side will show erythema only. Late mouth and throat dryness rarely occur, and the mucous membranes are moist in comparison with the glossy dry appearance of the mucosa in patients treated with parallel-opposed photon fields.

Fibrosis, severe in some patients, was seen when the electron beam was used alone in the treatment of lateralized lesions, like those of the buccal mucosa and retromolar trigone. Because of these unacceptable late radiation sequellae, the electron beam has been increasingly combined with other radiation modalities. In the patients so treated, fibrosis is uncommon. It is now the practice to combine the electron beam with 18- or 25-MeV photons (Fig. 3), ^{60}Co, or interstitial therapy in suitable situations, in order to lessen the high dose, not only to the skin, but also to the intervening tissues, and to avoid any underdosing, which might result from shadowing by bone.

CLINICAL APPLICATIONS

Since the capability of providing electron-beam therapy based on sophisticated dosimetry and techniques will not exist in all radiation therapy departments, it is important to define how indispensable the electron beam is in various clinical situations. It is infrequent that the electron beam offers the only possible modality that can provide the patient with the dose of irradiation that is needed. The major portion of electron-beam therapy is in the category either of offering specific advantages compared with the use of photons alone, or of providing an "elegant" method of treatment.

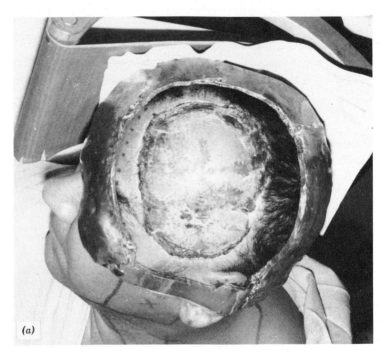

Figure 4. This 49-year-old male was seen at M.D. Anderson Hospital (MDAH) in December, 1973, with a history of having had a nodular mass in the left frontoparietal area of the scalp. There were four satellite nodules around the 3.5-cm mass. A wide excision of the area with split thickness grafting was done on December 4, 1973. The pathology report was angiosarcoma with numerous mitotic figures and vascular and lymphatic invasion. The patient was referred for radiation therapy because of the very aggressive nature of the tumor. Treatment was given with the 9-MeV electron beam to a large scalp field, 5000 rads given dose in 4 weeks. The parotid area was treated with combined 18-MeV photon and electron beams, receiving 5000 rads tumor dose at 4.5 cm in 5 weeks. The entire neck was treated with 9-MeV electrons, 5000 rads given dose in 4 weeks. Five months after completion of treatment, the patient developed a pleural effusion due to metastasis. He was placed on chemotherapy, which has controlled the thoracic disease, but in December, 1974, a small nodule on the scalp outside the irradiated area appeared, and, when tested, was positive for sarcoma. In May, 1975, there was no new evidence of disease. (a) Treatment field for the large grafted area in the scalp. Lead defines the field. (b) Parotid and neck fields. (c) Posterior extent of large neck field. (Reprinted by permission from Tapley, 1976)

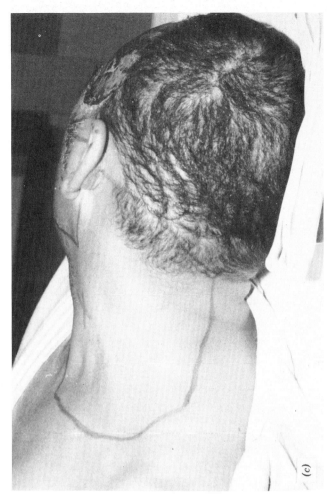

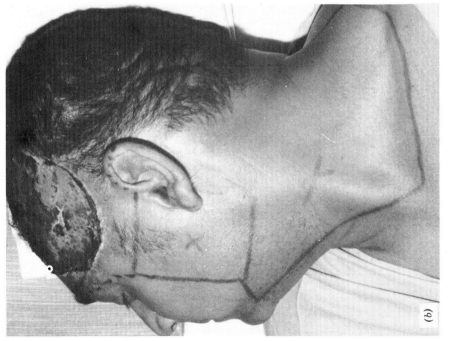

Fig. 4 (Continued)

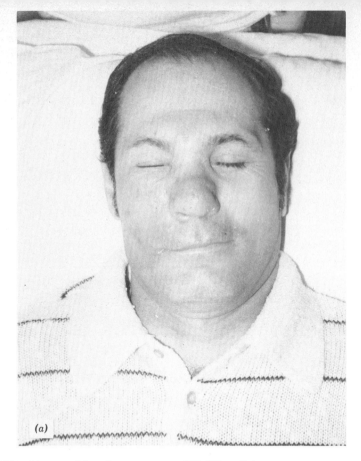

Figure 5. This 46-year-old male was seen at MDAH in July, 1973, with a history for the previous year of progressive thickening of the skin of the face, starting over the bridge of the nose. When seen at MDAH, the skin of the nose, glabella, nasolabial folds, right cheek, and upper lip were thickened and dusky in hue. The biopsy report stated, "This is an invasive carcinoma, which shows features suggesting basal cell carcinoma and others suggesting origin from hair follicles. The designation, skin adnexal carcinoma, is given because it cannot be categorized with certainty into a recognizable type of carcinoma." Treatment was given with 11 MeV, 7800 rads given dose in 47 days. The tumor dose at 2.5-3 cm was 7000 rads. The field was defined by a lead cut-out, and the eyes were partially protected by lead shields. Additional protection was provided with tungsten plugs. The field was not reduced at 5000 rads because of the diffuse nature of the disease. Photons could not be used because of the large treatment area and the marked irregularity of the field. There was heavy dry desquamation at the conclusion of treatment. During the 19 months since treatment, there has been slow disappearance of all thickening in the skin, except in the right cheek, where residual nodularity remains. The skin is gradually becoming pliable and will wrinkle on pressure. (a) The patient before treatment. Diffuse thickening of tissues of nose, right cheek and upper lip. (b) The patient in treatment position. The treatment field is defined by a cut-out in a lead mask. A mylar plate over the face holds tungsten plugs to protect the cornea and lens, because 11 MeV is being used. This would require 4.3 mm of lead to shield the cornea, and the eye shields placed under the lids measure only 2-2.5 mm. (c) The patient at conclusion of treatment showing dry desquamation over the entire field after receiving 7800 rads given dose in 47 days. (d) The patient at 14 months. There has been marked softening of the involved tissues. The nose and upper lip now appear relatively normal. (Reprinted by permission from Tapley, 1976)

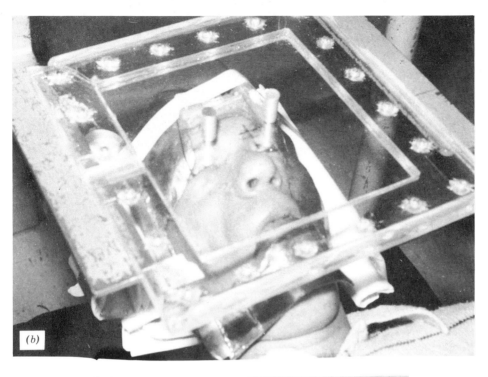

(b)

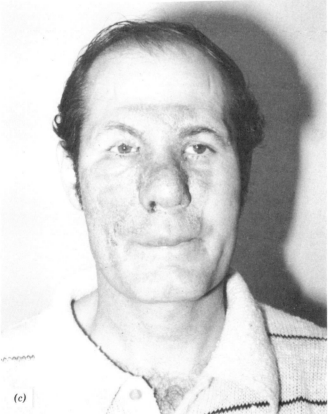

(c)

Fig. 5 (Continued)

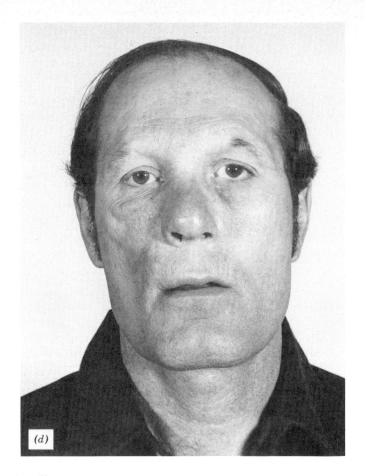

Fig. 5. (Continued)

NO ALTERNATIVE TREATMENT TO ELECTRON-BEAM THERAPY

There are a few clinical situations where the only possible method of irradiation is with the electron beam. In very large lesions involving the superficial tissues of the scalp and face (Figs. 4 and 5), there is no way to cover the total area that requires treatment with a photon beam, no matter how sophisticated the treatment planning.

ELECTRON-BEAM THERAPY IS A SUPERIOR MODALITY OF TREATMENT

In clinical situations where it is important to give a high dose to a limited volume, or to avoid overdosing structures deep to the tumor-bearing tissue, the electron beam clearly offers a superior method of therapy. Such clinical situations include: recurrent skin cancers around the eyes and nose, cancers of the nasal vestibule and columella, lateralized lesions of the buccal mucosa and retromolar trigone, unilateral neck treatment after radical neck dissection, treatment of the spinal accessory chain of nodes, and extensions of surgical scars which should be irradiated with the entire surgical area, but are difficult to include in the photon field.

Skin. A comparison of the dose distribution of 7- and 11-MeV electron beams with 100 kv, HVL-1.00 mm Al, and 140 kv, HVL-4.00 mm Al, irradiation is shown in Figure 6. To deliver a minimum dose at the desired depth with photons, a higher skin dose must be given. With 7, 9, and 11 MeV the surface dose is between 90-95% because of the rapid dose build-up and, therefore, a high dose level is maintained to a depth determined by the energy of the beam, with the dose decreasing sharply beyond this depth. For example, if a minimum dose of 5000 rads is selected at 1.5 cm, with 7 MeV the surface dose will be approximately 4800 rads, whereas with 140 kv, HVL-4 mm Al, the skin dose will be 6200 rads. With the electron beam, since the surface dose is only 5000 rads, the skin will remain in better condition years later.

Lesions which have recurred after several surgical procedures frequently involve the thickness of the subcutaneous tissues without gross involvement of the epidermis (Fig. 7). The tumor is often located in dense scar tissue with a poor blood supply and resultant tissue hypoxia. A high dose must be given over a longer treatment time to avoid excessive fibrosis. With all or a major portion of the treatment being given with the lower-energy electron beam, a high dose is maintained to limited depth.

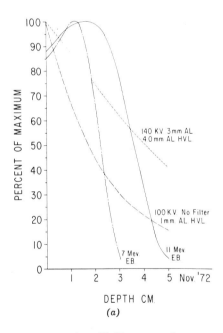

Figure 6. Depth-dose curves of 7- and 11-MeV electron beams, compared with depth-dose curves of "superficial" kilovoltage x-ray beams. (a) Central-axis depth-dose curves for 100 kv, HVL-1.00 mm Al, 140 kv, HVL-4.00 mm Al, 7-MeV electrons, and 11-MeV electrons. The disadvantage of the 140-kv beam is its gradual decrease in dose, at 4-cm depth being 50%, versus zero dosage for 7 MeV at this depth. The 11-MeV dose distribution resembles that of 140-kv irradiation, but the dose distribution is more homogeneous with 11 MeV, varying only 7% to a depth of 2.5 cm, with a rapid decrease in dose beyond this point. (b) Percentage depth doses for 7- and 11-MeV electron beams and for superficial radiation. (Reprinted by permission from Tapley, 1976)

DEPTH DOSES
ELECTRON BEAM – SUPERFICIAL RADIATION

Depth (cm)	7 Mev. E. B. (%)	11 Mev. E. B. (%)	100 K. V. HVL = 1.0 mm AL (%)	140 K. V. HVL = 4.0 mm AL (%)
0.5	91	93	82	94
1.0	99	98	67	88
1.5	95	100	55	81
2.0	67	99	45	75
2.5	30	93	39	68
3.0	3	78	30	62
3.5	–	56	25	56
4.0	–	32	21	50

Fig. 6. (Continued) (b)

Nasal Vestibule and Columella. Carcinomas of the nasal vestibule and columella present a complex treatment situation, requiring irradiation of deeper structures over a wide area, since the deep limits of the lesion along the septum or floor of the nose cannot always be determined. It may be necessary to treat up to the bridge of the nose or even higher to provide an adequate margin for the septum. Similarly the field must cover the floor of the nose and a portion of the upper lip, if the lesion developed in the vestibule. This area will receive relatively uniform dosage from the skin surface to the desired depth using an appositional 15- to 18-MeV electron beam. A portion of the treatment can be given with ^{60}Co, or an appositional electron-beam portal may be used directly over the columella and vestibule (Fig. 8). Either technique provides some sparing of the skin of the dorsum of the nose.

Because the nose is a midline structure, nodal involvement by disease may develop in either side of the neck. If this occurs, in connection with a neck dissection, treatment may be given with the 7-MeV electron beam to the lymphatic drainage pathways, extending from the nose to the submental tissues because of the possibility of "in transit" metastases (Fig. 9).

Lateralized Lesions of Oral Cavity. Wedged pair arrangements will spare the opposite side of the oral cavity in unilateralized lesions, such as those of the buccal mucosa or retromolar trigone, but the technique is fraught with complications. There can be a sharp gradient in dose between the minimal tumor dose and the dose to the mandible, but, alternatively, in order to obtain a minimal dose difference, and therefore not overdose the mandible, a long segment of the mandible has to be irradiated. Since complications are functions, not only of the dose and time, but of the volume irradiated, in both set-ups bone exposure and, at times, bone necrosis requiring resection will develop.

Although it is tempting to use the electron beam alone to treat carcinomas of the buccal mucosa, because these lesions are so well lateralized, previous experience has demonstrated an undesirable degree and duration of acute reactions, as well as later sequellae of severe skin changes and fibrosis with trismus. The current technique for buccal mucosa lesions combines the electron beam with interstitial gamma-ray therapy. The tongue is

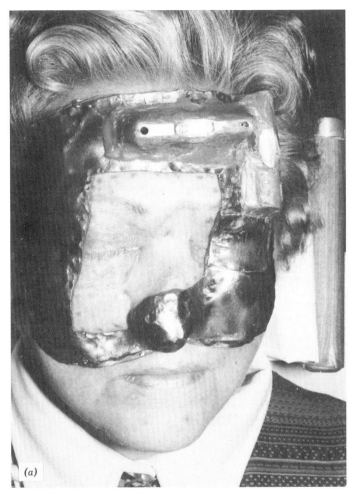

Figure 7. This 80-year-old female was seen at MDAH in April, 1974, with a long history of recurrent basal-cell carcinomas of the skin of the dorsum of the nose, extending into the frontal area. Multiple excisions and grafting had been done. The most recent procedure had been in March, 1974, when an area of thickening on the bridge of the nose was explored, and tumor was incompletely excised. On examination at MDAH there were areas of induration and nodularity, considered to represent tumor, associated with scars in the right nasolabial fold and the right medial supraorbital ridge. Treatment was given, using a lead mask to define the field, and an eye-shield to protect the cornea and lens. After 5750 rads given dose in 5 weeks to the large field, using 9 MeV, a dose of 500 rads in two fractions was given with 7 MeV to the area of thickening at the right medial supraorbital ridge. The treated skin showed brisk erythema and mild dry desquamation, which cleared completely at 6 weeks. At 9 months the patient had no problems, and all nodularity had disappeared. (a) Treatment field. Levels fixed to the lead mask provide accuracy in daily positioning of the head with reference to the vertical beam. (b) Patient 9 months later. Excellent appearance of operated and irradiated tissues. (Reprinted by permission from Tapley, 1976)

Fig. 7 (Continued)

protected by an intraoral stent containing lead. Usually 9 or 11 MeV is used, but if the lesion is very thick, 15, or even 18, MeV is necessary.

Carcinomas of the anterior tonsillar pillar (T_1-T_3, N_0-N_1), with minimal extension into the soft palate or tongue, receive combined treatment in a 1:1 or 1:2 ratio with the electron beam and 18- or 25-MeV photon irradiation, because a greater depth-dose is necessary, and because interposed bone will further decrease the depth-dose. Using the 18-MeV electron and photon beams and with equal given doses, the 80% tumor dose is located at a depth of approximately 6 cm (Fig. 3), unless shifted toward the surface by heavy cortical bone in the mandible. Corrections are made for mandible thickness, if necessary, when the tumor dose with the electron beam is calculated. The subdigastric node area is included in the treatment field of the primary lesion. The mucositis is sharply limited to the affected side, with essentially no mucosal reaction on the opposite side. Acute radiation morbidity is decreased, and late dryness is negligible. When there is a

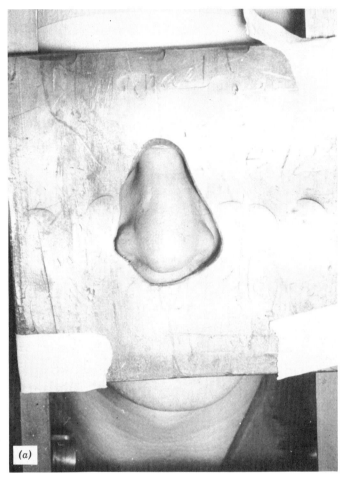

Figure 8. This 44-year-old male was seen at MDAH in October, 1968, with a 4-month history of a sore in the left nostril. A lesion had been removed from the left vestibule in May, 1968. Just prior to being seen at MDAH, the entire tip of the nose became very swollen, red, and painful. Surgical exploration revealed tumor involvement along the columella and septum, with extension into the left vestibule near the septum. All biopsies were positive for squamous-cell carcinoma, Grade II. Using 15 MeV, the entire nose was irradiated with 5000 rads given dose in 4 weeks through a lead cut-out, the beam being vertical. An additional 1650 rads was given to the columella of the nose with 9 MeV through a 3-cm cone appositional to the columella. At follow-up in April, 1975, the patient was doing well. The nasal cavity is kept clean and free of crusting by light application of mineral oil. There is a small asymptomatic perforation of the septum near the tip of the nose. (a) Lead cut-out. (b) Appositional field for columella. (c) Patient 5 years after treatment. (Reprinted by permission from Tapley, 1976)

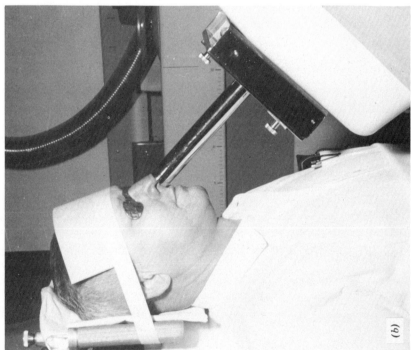

Figure 8 (Continued)

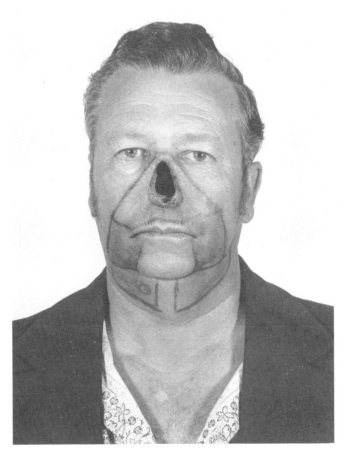

Figure 9. This 49-year-old male was seen at MDAH in January, 1975, with squamous-cell carcinoma of the left nasal cavity. On examination there was reddish discoloration of the tip of the nose and the left ala nasi. There was an ulcerated lesion involving the anterior third of the nasal septum, extending posteriorly on the left. The columella was indurated. A total excision of the nose, septum, and nasal concha was done. The pathology report described tumor present at the left upper septum margin, and the patient was referred for radiation therapy. Through a field defined with a lead cut-out, 5000 rads given dose in 4 weeks was delivered with 9 MeV to the operated area. Through a 3-cm circle, using 9 MeV, an additional 1000 rads in four fractions was given to the area of residual tumor in the septum. In 4 weeks 5000 rads given dose was delivered with 7 MeV to the lymphatics of the skin bilaterally from the base of the nose down to the submental area. With an anterior ^{60}Co portal, using a narrow midline block, 5000 rads given dose in 5 weeks was delivered to the upper neck. (Reprinted by permission from Tapley, 1976)

clinically positive subdigastric node, treatment is given to the entire ipsilateral neck with the electron beam. A neck dissection must be done if the dose to the nodes is not higher than 5000 rads.

Elective Irradiation of Unilateral Neck. For postoperative irradiation of a radically dissected neck, one cannot use anterior and posterior portals with a photon beam. If parallel-opposed portals are used, the mucous membranes of the mouth and throat are heavily irradiated. With the electron beam, unilateral irradiation after a radical neck dissection can be done with minimal irradiation to the deep structures.

Indications for treatment of one entire neck after radical neck dissection include:

1. Many positive nodes in the surgical specimen,
2. Total replacement of a node by tumor or rupture of a cystic node,
3. Tumor found in connective tissue (Fig. 10), and
4. Tumor found in the perineural lymphatics.

Healing must be complete before irradiation is started, and the treatment field must include all extensions of the surgical scar. This can present difficulties, particularly when the anterior limb of the neck incision extends directly across the thyroid cartilage. It is then necessary for the larynx to remain in the field without shielding until 4000 to 4500 rads have been given. If possible, 7 MeV is used, which will limit the dose to the ipsilateral laryngeal structures. If the scar crosses the midline into the opposite neck and if there is a high risk of contamination of this portion of the surgical wound, a small appositional field is used to cover this scar.

Spinal Accessory-Chain Nodes. Irradiation of the spinal accessory chain nodes is indicated in a variety of situations, including primary lesions in the nasopharynx and tonsillar fossa, and whenever there are large, clinically positive, subdigastric nodes. If the primary lesion is treated with parallel-opposing fields, using a 1:1 loading, the fields are extended back over the accessory chain of nodes until the midline tumor dose reaches 4500 rads. To exclude the spinal cord from further irradiation, the posterior margin of the primary field is moved forward, and 7 or 9 MeV is used to treat the posterior strips to at least 5000 rads at a 1-cm depth. When parallel-opposing fields with a 2:1 loading are being used to treat the primary lesion, a posterior strip on the contralateral side may be treated entirely by 9 MeV and by the exit dose of the ipsilateral field, which extends back over the spinal cord. In the treatment of a carcinoma of the nasopharynx which is unusually complex, the electron beam may be used to treat both posterior strips to 5000 rads (Fig. 11).

Extensions of Surgical Scars. After surgical procedures where there is a high risk of infestation of the operated area, no part of it should be left unirradiated. For example, surgical scars extending up to and over the mastoid tip should be irradiated. This is difficult with the photon beam, whereas these areas are easily treated with the low-energy electron beam.

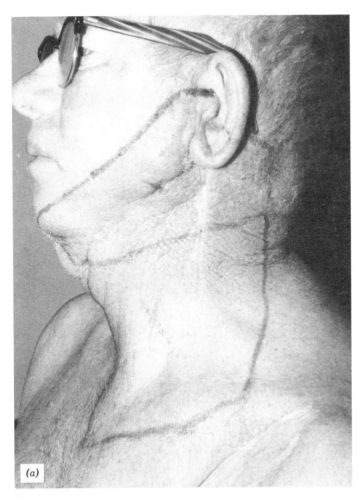

Figure 10. This 57-year-old male was first seen at MDAH in March, 1974. A small carcinoma of the left lower lip had been cauterized 9 months before. Six months later he developed a mass in the left submandibular area, which grew slowly and invaded the overlying skin. When seen at MDAH, the mass measured 4.5 cm and was partially cystic and fixed to the overlying skin. In March, 1974, the patient had excision of the lower lip scar, resection of the left submandibular mass and overlying skin, a modified left-neck dissection and a superficial parotidectomy. Histopathology was metastatic squamous carcinoma in submandibular lymph node and connective tissue, 31 negative lymph nodes, negative parotid and submaxillary salivary glands and negative lip scar. Radiation therapy was given to the entire left neck. The upper neck was treated with 9 MeV, the lower neck with 7 MeV; 5000 rads given dose was received in 4 weeks. The patient had no evidence of disease in December, 1974. (Reprinted by permission from Tapley, 1976)

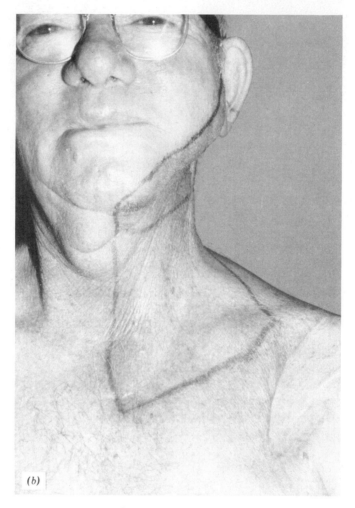

Figure 10 (continued)

ELECTRON BEAM THERAPY HAS AN ADVANTAGE OVER OTHER TREATMENT MODALITIES

There are clinical situations which can be satisfactorily treated with the photon beam, but where the electron beam may offer a somewhat more "elegant" technique. An advantage is provided through ease of treatment and assurance that the dose to underlying structures will be minimal. Such clinical situations include treatment to the parotid bed, boost therapy to lymph nodes, and treatment to the chest wall after radical mastectomy.

Parotid Bed. Treatment to the parotid area can be given with a ^{60}Co wedged pair, but the electron beam, usually in combination with 18 to 25 MeV photons, is the treatment of choice in situations when the bulk of the tumor has been removed and the 7th nerve is not grossly involved. One is dealing only with possible microscopic disease in these selected clinical situations, which include:

1. Total tumor removal with close surgical margins,
2. Total gross tumor removal with tumor cut-through demonstrated by pathologist,
3. Total tumor removal with histopathologically high-grade malignancy or metastic squamous cancer from skin, and
4. Total removal of recurrent tumor after total parotidectomy, with recurrence limited to small skin nodule(s) and no evidence of deep invasion.

The area to be irradiated includes the entire parotid bed and the full extent of the surgical scar. The total irradiated volume is limited, as compared to the ^{60}Co technique, but excellent coverage of the parotid fossa is provided (Fig. 12). With combined electron and photon beams, a relatively homogeneous dose distribution is achieved from the surface to a 5-cm depth. Since the posterior extent of the field usually covers the spinal cord, the field is moved off the cord at 4500 rads. Seven or 9 MeV is then used to give 1000-1500 rads to the posterior portion of the field, particularly to include the surgical scar when it extends behind the ear.

"Boost" Therapy to Lymph Nodes. Boost irradiation to nodes can be given with photons through glancing fields, thus avoiding the deeper structures, but a satisfactory technique is provided by the electron beam using appositional fields. After the area of the primary and both necks have reached 5000 rads with ^{60}Co, the electron beam is used at varying energies to increase the dose to involved nodes in the neck or to areas of high risk of infestation, without increasing the dose significantly to the deeper structures, such as the pharynx, larynx, or spinal cord. The additional dosage is given through a field limited to the area of residual adenopathy or induration. The electron beam is also used to treat a field-within-a-field, if a lymph node continues to enlarge under treatment.

Chest Wall After Radical Mastectomy. The electron beam is preferred for the irradiation of the peripheral lymphatics areas and the chest wall after radical mastectomy because of the ease of treatment and the minimal lung irradiation. The supraclavicular nodes are located just under the skin and could be undertreated unless build-up material is used, if irradiated with the 4- to 6-MeV photon beam. Treatment of the internal mammary node chain with the megavoltage photon beam gives a high dose to the mediastinal structures, including the hilar vessels, pericardium and myocardium. With the electron beam, the energy is selected for the depth of minimum tumor dose, with rapid fall-off in dose beyond this point. The supraclavicular area is treated with 11 MeV, which provides the minimum tumor dose at 3-cm depth, and the internal mammary node chain is treated with 15 MeV to provide the minimum tumor dose at 4-cm depth.

The use of the 7-MeV, straight-on electron beam is an elegant solution for a treatment of the chest wall which avoids some of the dose to the lung produced by tangential ^{60}Co fields. The average chest-wall thickness after radical mastectomy is 1.5-2.0 cm, a depth well covered by the 90% dose range of 7 MeV. When the chest-wall thickness approaches 3 cm, 9 MeV is used. The chest-wall portal must include the full extent of the mastectomy scar, but if the scar cannot be satisfactorily covered by one field, small contiguous fields are added.

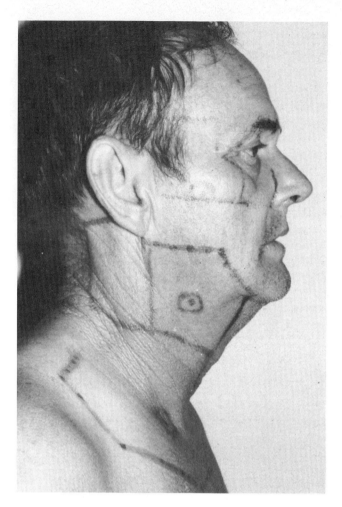

Figure 11. This 52-year-old male was seen at MDAH in September, 1974, with the diagnosis of poorly differentiated squamous-cell carcinoma of the nasopharynx extending into the posterior nasal cavity. There was invasion of the ethmoids and the left maxillary antrum. There were bilateral midjugular nodes, most numerous on the left. The patient's disease was staged T_4N_{3B}. The patient received radiation therapy from September to December, 1974. The nasopharynx received 6800 rads tumor dose in 12 weeks with an anterior open field and bilateral wedged ^{60}Co portals. Bilateral upper-neck fields received 6000 rads tumor dose with parallel-opposed ^{60}Co portals, and the lower neck received 5000 rads given dose with an anterior ^{60}Co field. Both spinal accessory-chain nodes were treated with 9 MeV and received 6000 rads given dose in 9 weeks, 250 rads per fraction. Treatment was interrupted from the end of the fourth week until the beginning of the eighth week because of weight loss and dehydration due to swallowing difficulties associated with the intense mucosal reaction. When treatment was resumed, the patient's condition had improved, and he tolerated the completion of the planned course of therapy relatively well. A lateral view of the patient's treatment fields is shown. The electron-beam field for treatment of the posterior cervical chain nodes is placed posterior to the upper neck ^{60}Co field and extends from the mastoid tip down to the line which defines the upper margin of the anterior lower-neck field. (Reprinted by permission from Tapley, 1976)

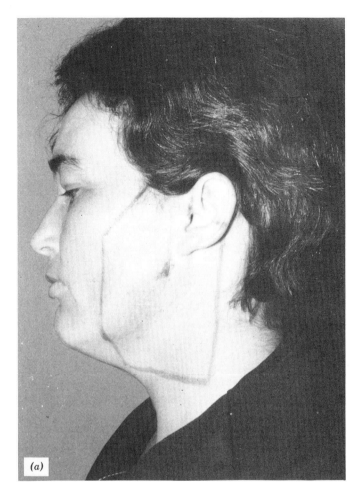

Figure 12. This 35-year-old female was seen at MDAH in June, 1974, having had a recent resection of a parotid tumor with preservation of the facial nerve. She had a 5-year history of a slowly growing mass in the left preauricular area, which 6 months prior to admission began to grow rapidly and became tender. In May, 1974, a superficial parotidectomy was done, and the histopathology of the tumor was acinic-cell carcinoma. The outside surgeon stated that the lesion appeared to be encapsulated but was directly overlying the 7th nerve. In order to obtain a good margin, it would have been necessary to sacrifice the nerve. The patient was referred for radiation therapy because of the possibility of residual tumor. Treatment was given to the parotid gland area through a single, left, lateral portal with 18-MeV electron and photon beams, 5500 rads tumor dose at a depth of 5 cm in 5½ weeks. The loading was 2:1 in favor of electrons. The subdigastric nodes were included in the field. At 4500 rads tumor dose the field was decreased posteriorly to exclude the spinal cord. The posterior strip area received an additional 1000 rads with 9 MeV. It was not necessary to treat the entire neck with this low-grade tumor. (a) The treatment field extends above the temporomandibular joint, includes the subdigastric nodes, and extends back to the mastoid. The upper half of the pinna is blocked with lead. (b) The posterior portion of the field has been excluded from the combined electron and photon beams and is being treated with 9 MeV. (Reprinted by permission from Tapley, 1976)

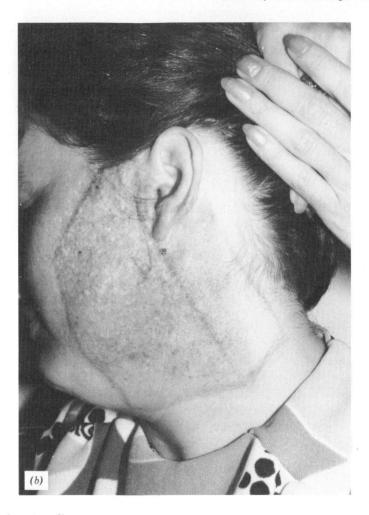

Figure 12 (continued)

REFERENCES

Almond, P.R. 1976. Radiation physics of electron beams. *In* Tapley, N. duV. (ed.) Clinical Applications of the Electron Beam. John Wiley & Sons, New York.

Almond, P.R., A.E. Wright & M.L.M. Boone. 1967. High energy electron dose perturbations in regions of tissue heterogeneity. *Radiology* 88:1146–1153.

Boone, M.L.M., E.H. Crosby & R.J. Shalek. 1965. Skin reactions and tissue heterogeneity in electron beam therapy. II. *In vivo* dosimetry. *Radiology* 84:817–822.

Boone, M.L.M., J.H. Jardine, A.E. Wright & N. Tapley. 1967. High energy dose perturbations in regions of tissue heterogeneity. I. *In vivo* dosimetry. *Radiology* 88:1136–1145.

de Almeida, C. & P.R. Almond. 1974. Comparison of electron beams from the Siemens betatron and the Sagittaire linear accelerator. *Radiology* 111:439–445.

Fletcher, G.H. 1973. Textbook of Radiotherapy. 2nd Edition, Lea and Febiger, Philadelphia.

Sinclair, W.K. & H.I. Kohn. 1964. The relative biological effectiveness of high energy photons and electrons. *Radiology* 82:800–815.

Tapley, N. duV. 1976. Clinical Applications of the Electron Beam. John Wiley & Sons, New York.

Tapley, N. duV. & G.H. Fletcher. 1965. Skin reactions and tissue heterogeneity in electron beam therapy. I. Clinical experience. *Radiology* 84:812–816.

Discussion

Dr. Tapley: I have a question here from Dr. Munzenrider: "Could you comment on the use of electron-beam whole-skin therapy in *Mycosis fungoides*?" The electron beam offers a splendid way to treat the entire skin of the body. If one uses a sufficiently low energy, the depth-dose is minimal. *Mycosis fungoides* is a very complex disease; whole-skin therapy should be considered only at certain phases, when the lesions are relatively thin. At the stage where the tumors may be as thick as 2 cm, you would have a problem using the equivalent of 2- to 3-MeV electrons. The dosimetry is very complex and has to be worked out very carefully ahead of time. All skin areas must be covered, with some protection around the eyes, unless the eyelids are involved. There are problems, but it is an ideal treatment if you have the right clinical situation and your technique properly worked out.

Dr. J. Dutreix: I have a question: "What is the survival of the esophageal cancer after split course treatment?" I do not have the exact figures yet, but the average survival for cancer of the esophagus is 11 months for conventional fractionation, as well as when the split course is steady. The survival after six months is about 50%, and after 18 months something like 15%. We are not convinced that we have cured these tumors except in a few cases. One of the first patients treated on our Allis-Chalmers betatron, in 1953, was a case of cancer of the esophagus. The patient died three months later; the tumor was sterilized. The patient died from pulmonary complications, from liver metastasis and so on. We were very excited at the time at having proof of the possibility of curing cancer of the esophagus with the betatron. I am afraid that has been the only one we have cured locally.

Dr. Levan: Are you using non-standard fractionation at the Institute Gustave-Roussy? If so, are you using concentrated irradiation and multiple fractions?

Dr. Dutreix: Yes, we use concentrated irradiations extensively—two times, 850 rads at two-day intervals for palliative and preoperative irradiation; we also use this as the first part of treatment with curative intent. In split-course treatment, we use a second series which is usually fractionated. For brain tumors we use two concentrated irradiations after one-month intervals, two times, 850 rads, or two series of three fractions of 600 rads.

Discussant: If I recall correctly, one of your colleagues, Dr. Ennuyer of Villejuif, advocated extensive rather than concentrated irradiation, i.e., treating six or seven days a week. Would you care to comment on this technique?

Dr. Dutreix: Dr. Ennuyer is a radiotherapist of the Foundation Curie, where they are more inclined to go to high fractionation (started by Baclesse a long time ago). Ennuyer also advocated extended, rather than concentrated, irradiation.

Dr. Almond: There are a number of questions concerning collimation and the differences between the linear accelerator and the betatron, which I showed in my presentation. Dr. Pohlit questioned at that time (when the slide was shown), indicating he thought I labelled too much collimation; the ears on the edges of the beam would lead to better distribution at depth. It certainly flattens out. It was our feeling that those hot spots are quite hot, and would not be acceptable clinically. When you have the combined fields Dr. Tapley showed for chest-wall treatments and other areas, these hot spots would add up to much hotter areas than we currently get.

Dr. Johns raised further questions concerning the differences between the Siemens and the Sagittaire linear accelerator. These machines are quite different. The Siemens is a betatron; the beam comes out in a circular fashion through a different collimating system. It has a treatment distance of 50 cm. The Sagittaire is a linear accelerator, going through a magnetic system; treatment is at about 92 cm from the target. The primary explanation for the differing shapes of depth-dose curves is that one machine has scattering foils in the beam. The collimation will affect the shape of the isodose curve, but not the central-axis depth-dose.

Dr. A. Dutreix: I don't agree. I think the collimator very much affects the depth-dose scale.

Dr. Bloedorn: We constructed a collimator similar to the one you have and we couldn't detect any differences in any depth-dose distribution.

Dr. Almond: It's a very complicated subject. There was a question about the material for the collimators. Our experience is that you need some high-Z material. We did some work with lucite on the Allis-Chalmers, and we were not really satisfied with it. There are data in the literature to show that, close to the patients, you may need combinations of material, a sandwich, to get the desired distribution. It's very complicated. The collimators, scanning mechanisms and other things will have an effect on the differences in the distribution. I only mentioned periodic measurement of the dose output. Certainly, energy checks should be done on a regular basis. Dr. Pohlit has reported in the literature some very nice devices for doing this, for example, ratios of measurements at different depths. There was a question here: "Do you think that electrode or photo disintegration have a place in energy measurements for electrons?" That's one method of doing it, but it's a long, tedious method; there are quicker and easier methods for routine use. One should check the homogeneity of the beam regularly, and other parameters that can vary should be checked on a regular basis.

Discussant: Dr. Watson, of the clinical trials in radiotherapy, almost the only one to show a significant difference was the Allt series on cancer of the cervix at Princess Margaret Hospital. The 5- and 10-year survival was double for the betatron. Because a cell doesn't really care what machine produces the ions, why is the betatron superior in Allt's series?

Dr. Watson: The only explanation is the difference in physical distribution of the high-dose area. If a clinical trial shows that, then it has to be. People who have used the betatron in late cancer of the cervix—Dr. Fletcher, for example— immediately see some tremendous difference in their results.

Dr. J. Dutreix: Professor Jack Fowler asks, "How did you arrive at the interval of two hours between your two doses of 180 rads each? Was it based on the recovery peak of cells *in vitro*? We have found that for skin, *in vivo*, a much longer interval, at least three hours, is necessary for full repair. It depends strongly on the size of dose, too, which is not the case with cells in culture." We decided to use the two-hour interval after some experiments on human and mouse skin, which showed D_2 minus D_1, the additional dose, when the single dose is split into two parts, increased rapidly during the first 2 hours and then leveled off between 2 and 3 hours. While we do not have full cellular repair at 2 hours, it is close to full. The smallest possible interval compatible with full repair was chosen just for the out-patients' convenience.

Dr. Fowler: Would it be better to explore different fractionation with fewer and larger fractions or a larger number of small fractions?

Dr. J. Dutreix: As we saw yesterday in Eric Hall's presentation, the cell survival curves for all the biological materials tested seem to have some initial slope. So for small doses, around 200 rads, we are on the initial tangent; the cell killing for such doses is due mainly to lethal, and not to the accumulation of sublethal, damage. Therefore it doesn't matter what the fractionation is. If you go to smaller fractions you cannot expect to effect an important change, as long as the overall time is kept constant to keep the repopulation constant. On the other hand, if you use larger fractions, you enter the range of doses where you have some cell killing due to accumulation of sublethal damage. This may differ from one species to another. I would expect more different transient effects between two tissues when using fractions larger than the 200 rads than those smaller. That is just a rational guess; it has not been demonstrated.

Dr. Tapley: I have a question concerning optimal beam energy: "It appears that the 6- to 18-MeV electron energy range is most often used." I have used it most often because that's what I had available. However, we are now using 19 MeV electrons with which we do get a greater depth coverage; we can use it for lesions 6-cm deep. The question continues, "Do you advise that a department have electron capabilities beyond 18 MeV? If yes, what higher limit would you suggest?" I would rather depend on the depth of the tumor than the energy level, which differs with different pieces of equipment. With the particular characteristics of the electron beam, clearly advantageous for superficial or lateralized lesions, beam energies higher than those giving a satisfactory depth-dose at around 6-7 cm would be of no particular advantage. When one has tumors located beyond this depth, the therapeutic ratio should be considerably better with high-energy photons. I would suggest, as did Dr. Ho, an upper limit for electrons of 25 MeV.

Dr. J. Dutreix: We have available electrons up to 32 MeV, and I agree entirely. Beyond 7-cm depth photons usually give a much better dose distribution than electrons. There are a few special uses for electrons of energies above 20 MeV, or reaching beyond a 7-cm depth. We have found electrons very convenient for irradiation of the spine in bone metastases of the spinal column. For the spine, in adults, it's sometimes necessary to go up to 32 MeV, but 90% of our treatments are done under 20 MeV. I agree that a 7-cm depth is the best rule for the upper limit of energy. But when higher energy is available, one sometimes has a use for it.

Dr. Tapley: Dr. Ho asks, "In treatment of neck nodes with electron beam by direct lateral field, do you encounter arytenoid edema?" Very rarely. Actually the patient I discussed with the carotid field and the whole neck irradiated did develop some difficulties at around the 4000-rad level. His problem was mainly one of mucositis along the lateral laryngeal wall, but he was beginning to develop a little laryngeal edema. Actually we started using 11-MeV electrons for his treatment because his tissues seemed to be so thick. We then decreased the energy, and the beginning edema subsided. If edema does begin to develop, then the energy should be lower for that patient. Another question concerns the chest-wall treatment, and treating the line twice. I mentioned the hot spots in the skin and that they healed quickly. It is pointed out by Dr. Chu that hot spots in the skin may be acceptable because the brisk reaction subsides; however, she thinks you get problems with rib fractures due to hot spots at the level of the ribs. Actually, we have not been seeing rib fractures. We did have a few in the beginning with the electron beam when we were giving 6000 rads, given dose, in four weeks to the chest wall, and the patient was rotated up on his side to flatten out the chest wall. That meant that the dose throughout the rib cage was high. We stopped that, however, mostly because of pneumonitis and crossfire in the mediastinum. We also were using 9 MeV more often in those earlier days. I had rib fractures in three or four of those patients, but of the 500 chest walls we have treated, we haven't seen any in the last few years, and I have not as yet seen them in the patients where we've been treating the lines as we have.

Dr. Laughlin: Dr. Webster asks: "Would you comment further on the practical implementation of computerized axial transverse tomography (CAT) in current treatment planning? It would appear that relative absorption coefficients, for example, could be readily interfaced for inhomogeneity corrections, dose planning, and so on. Also, is it practical cost-wise? Much of this could be done with computer software." We think it is of great practical importance for treatment-planning purposes. In the case of implants, AP and lateral radiographs already provide very good localization, and by doing two-dimensional analysis on each of these, the computer program will give you the three-dimensional coordinate. While it isn't necessary for implants for the seed localization per se, the CAT scanners have the advantage of relating this seed location more clearly to anatomy than is possible with the present stereo shift method or the AP and lateral radiograph analysis. For external-beam treatment it appears clearly superior, at first glance anyway, to what one can do now with transverse film tomography. With regard to cost, the practicality remains to be seen, but I should point out that it isn't necessary for therapy units to have the resolution required in the diagnostic units. We have designed a cesium-137 unit which, unfortunately, is not funded. This was presented at the meeting of the Radiological Society of North America a couple of years ago. It is entirely feasible to build a unit completely adequate for therapy planning for less than \$100,000. Another method, to accomplish the same purpose, is to use your cobalt-60 rotation unit, put a fan-geometry insert on it, and use film rather than detectors to integrate. Then you will have to densitometer the film, and use an algorithm in the same way that you would use for a detector. That method would take longer. I am convinced that tomography will become very important in treatment planning in the future by one or more of these different approaches.

Dr. Almond: I've tried to group some of these questions; some are mainly concerning the C_E factors for low energy. The empirical relationship that I gave you is good over the energy range from 7 to 20 MeV, the energies that we have at our institution. Comparing values derived by that relationship with the values in ICRU-21 at the correct energies and depth, we're off by no more than 1% at any particular point. That relationship does break down at the lower energies. At the end of the range where your lower energy is, the uncertainty in the C_E is not as important, and so even if there is a 3, 4, or 5% uncertainty in the C_E, it will not change the shape of the depth-dose curve to any noticeable extent. We have to be careful, however, when we go to the initial electron-beam energies of 4 or 3 MeV that people are now using. Much more work needs to be done to measure not only the C_E factors, but also perturbation and displacement corrections at that energy; there are no data available in that range at the present time.

There were a number of questions about buildup caps, asking what I was referring to. With the standard lucite buildup cap, the attenuation factor is 0.985 for cobalt-60. I have obtained recently a buildup cap of air-equivalent plastic, and its attenuation factor was 2% lower than that and did, in fact, make that much difference when I used that particular chamber for electron beam calibration. One must be careful not to take all of these factors completely for granted without looking at all the possible variations. When doing measurements in electron beams, the buildup cap should be taken off to leave as thin a wall as possible.

Dr. Watson: I have a question about the 35-MeV betatrons. It says: "With a 35-MeV betatron would you recommend setting the x-ray energy to, say, 25 MeV if the dose rate is sufficient, or would you prefer 35 MeV for better dose distribution?" The dose distribution is a little better at 35 MeV than at 25. From our 35-MeV betatron, a Brown-Boveri, the output with reasonable flattening of the beam for a large-sized field is only 40 rads per minute. If this machine were set lower in energy, the dose rate would be even worse than it is.

Dr. Tapley: I have a question: "Have you done any comparative studies to evaluate high-energy beams against other possible beams?" I will assume for the purpose of answering that it asks whether we have done controlled or comparative studies of the electron beam and cobalt or 25-MeV photons. We have not. We haven't thought of the electron beam as competitive with the photon beam, but rather as complementary to the other radiation modalities.

Dr. J. Dutreix: We haven't made any random studies between conventional and high-energy radiation; there is not much use in comparing two series which are distant in time. We have done clinical trials comparing high-energy photons and electrons from the Allis-Chalmers betatron. We started this trial in 1967; we stopped after treating about 20 patients with x-rays and 20-25 patients with electrons. The statisticians told us the results were already very convincing. We were looking at the tumor regression, which was much greater with the x-rays than with the electrons; we concluded that the final results would also be better with x-rays. At that time the physicists at our institution, going back to their own estimation of the dose, found out that they were in error by about 7% in the estimation of the dose given by the electrons. Further, there was some additional variation, because there is a kind of a priority in the use of photons on the betatron, and when there was any alteration in the number of fractions per week, it was always with the electrons and never the photons. When we took into account the average number of fractions per week, it turned out that the difference we were observing was perfectly compatible with an equal RBE on both beams, and that this difference was related only to the time-dose factors.

Dr. Dutreix: Dr. Perez asks about electron absorption in bone. We cannot handle the absorption in bone in the same way with electrons as we do with photons. Encountering a bone layer, we advance the depth of the dose for electrons. The dose does not change beyond the bone. The total distribution is just a push forwards. If you express the depth in terms of grams per square centimenter, the distribution is not affected by the bone itself. At 22 MeV, the electrons gave 80% depth dose at a 6.5-cm depth. We always reached the half line. The dose the tumor received was the prescribed dose.

Dr. Tapley: If you have an 18-MeV betatron electron beam, you do have to be aware of the mandible and increase the dose, usually about 10%, if it's a heavy mandible.

Dr. Laughlin: Relative to skin-dose, a question has been posed to Dr. Powers asking what he considers the significant depth for skin reactions.

Dr. Powers: The primary tissue of concern is the basal cell. The second is the immediate subcutaneous tissue. But a calculation has to go from the surface all the way down because one is dealing with tangential beams. We have had evidence presented that there is a difference between the Siemens betatron and Sagittaire linear accelerator as far as electron beams are concerned. As Dr. Almond has reported, there are changes in depth of D_{max} and surface dose with changing fields in very high-energy beams from accelerators; I suspect that if one got large enough beams from betatrons, one would have exactly the same phenomenon. I think the surface dose is dependent on the field size, the energy, the radiation unit, and very probably whether or not the beam is tangential. The computational methods become extremely difficult in that set of circumstances.

Dr. Almond: There is a question about the difference between the curve Dr. Watson showed with the cobalt 60 and the 34-MeV x-ray beam profile, as compared to the Sagittaire's beam. It's about the same on the Sagittaire. All high-energy x-ray beams have a penumbra of 1 to 2 cm; I don't think you can get rid of it.

Discussant: But Dr. Watson showed that at 4.5 cm it was 90%.

Dr. Almond: I don't think it was quite that wide and ours is probably 2 to 3, but they are there.

Mme. Dutreix wanted to know about recombination in commerical chambers with electron beams. We have found it necessary to increase the voltage on commercial chambers. We're not satisfied unless we have 300 volts on a Farmer chamber, for example, so that at several hundred rads a minute of electron beams, they are not saturating.

Mr. Strubler asks: "What is the physical explanation for the increase in surface dose with increase in electron energy?" Dr. Pohlit has the best explanation. It has to do with the characteristics of the electron beam, so it should be true of all accelerators. If you look at the theoretical calculations of people like Spencer, many, many years ago, and Martin Berger with his Monte Carlo calculation techniques, it is very true that the surface dose increases as the energy increases. It has to do with the scattering characteristics of electrons at different energies, and it is opposite to the x-ray curve.

Problems of High-Energy X-Ray Beam Dosimetry

Andrée Dutreix, Ph.D.,
Chief of Radiophysics,
Institût Gustave Roussy,
Villejuif, France

High-energy x-ray beams have been used in radiotherapy for more than 20 years, and many papers have been published on their dosimetry. However, the difficulties encountered with their use are too often underestimated. Dose measurements determined with an ionization chamber, especially in regions with high dose gradient, are one problem. Another, which has not received sufficient attention, is the importance of the factors affecting the shape of depth-dose curves and the need for a better specification of the beam quality. Finally, the clinical implications of the dosimetric characteristics of high-energy x-ray beams, such as the depth of the maximum absorbed dose, the sharp edges of the beams, and the relatively slow attenuation of the x-rays, need to be considered.

DOSE MEASUREMENTS WITH IONIZATION CHAMBERS

Relative depth-dose can be measured with various detectors. Films, thermoluminescent dosimeters, and flat ionization chambers have been used and compared for a 25-MV x-ray beam. Similar results are obtained with these different detectors if the dose is assumed to be measured at the depth of the front wall in the flat ionization chamber.

However, the problem is more complex for cylindrical cavities. The ionization in the air cavity is produced by the interaction of the secondary electrons. As the absorbed dose is proportional to the electron fluence, the air ionization has to be related to the electron fluence at a point in the phantom material in absence of the cavity (Fig. 1).

This point P, called "physical center of measurement," is distinct from the geometrical center 0 of the cavity for high-energy x-rays. Assuming the electrons are emitted strictly forward, and disregarding the influence of the central electrode, a theoretical calculation indicates that the distance a between the physical center and the geometrical center is equal to $0.85\,r$ where r is the radius of the cavity.

Recent measurements have been performed in two high-energy photon beams (5.5 MV and 25 MV x-rays). A cylindrical chamber (6 mm in diameter by 10 mm in length) has been compared with a flat chamber (1.5 mm in thickness by 5 mm in diameter). The relative readings of the two chambers were plotted (Fig. 2) as a function of the depth of the front wall, for the flat chamber, and of the depth of the geometrical center, for the cylindrical chamber. Measurements in the build-up region lead to the determination of the physical center of the cylindrical chamber which has been used. Assuming the

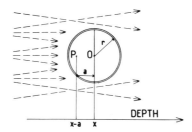

Figure 1. *The reading of a cylindrical ionization chamber has to be referred to the absorbed dose at the physical center P and not at the geometrical center 0.*

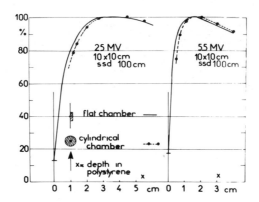

Figure 2. Comparison between relative readings of two ionization chambers, as a function of the depth x of their geometrical centers.

physical center of measurement in the flat chamber to be on the front wall, the physical center of measurement in the cylindrical chamber appears to be in front of the geometrical center at a distance $a=0.85\ r$ equal to the theoretical value. These results are in good agreement with the data from Hettinger et al. (1967) and with our own data for electron beams (Dutreix and Dutreix, 1966). At points deeper than the maximum, the determination of the position of the physical center is more difficult since a depth variation equal to a leads to a dose variation as small as 1 or 2%, for $a=2.5$ mm. For larger cavities this error may have a clinical significance. The correction for the actual position of the physical center determined on the beam axis can probably be applied to off-axis points, except near the edges of the beam, on account of the asymmetric distribution of the electrons.

The amount of electrons emitted at a large angle is not negligible, although the emission of secondary electrons is mainly in the forward direction, and the conditions of lateral electronic equilibrium are not fulfilled near the beam edges. A clear demonstration of non-lateral electronic equilibrium can be achieved by measuring the build-up curves for very small field sizes (Fig. 3; Dutreix et al., 1965). A beam diameter equal to or larger than 30 mm is necessary to assure full lateral electronic equilibrium.

As there is a lack of electronic equilibrium near the beam edges, for 25 MV x-rays at distances shorter than 15 mm of the geometrical limit, the distribution of secondary electrons is not symmetrical, and there should be a lateral displacement of the physical center of measurement in cavity towards the beam axis.

In such regions of transition, the dose distribution is assessed more accurately with film dosimetry than with air ionization measurements.

FACTORS AFFECTING THE SHAPE OF DEPTH-DOSE CURVES

Variation of transition curves. In the first centimeters of tissue the dose is increasing from a minimum value at the skin to a maximum at a depth depending on the x-ray energy and the geometrical parameters.

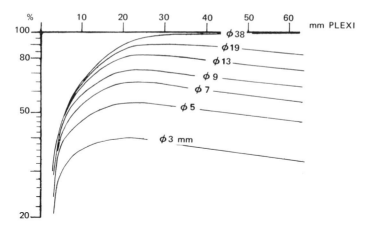

Figure 3. Depth-dose curves for small-diameter beams of 25 MV x-rays. The beam diameters vary from 3 to 40 mm. This set of curves shows clearly the need to enlarge the beam diameter to at least 40 mm to achieve full lateral electronic equilibrium.

If there is a vacuum between source and patient, the skin absorbed dose should be very small because the number of electrons back-scattered by tissues is small. In practical conditions, the contribution to the skin dose of electrons originating from air is small, too, since the beam diameter and the SSD are considerably smaller than the electron ranges in air, and thus are never large enough to assure either longitudinal or lateral electronic equilibrium (40 meters of air would be necessary for 20 MV x-rays).

However, some electrons originating from the collimator or the various beam modifiers placed in the beam can reach the skin. They increase the skin dose and reduce the depth of maximum absorbed dose. Figure 4 shows the variation of transition curves measured in a water phantom when a 3-cm perspex plate is moved in a 20 MV x-ray beam from 0 to 40 cm from the surface. Modifications of depth dose curves are observed at depths greater than the depth of the maximum absorbed dose, since the maximum range of secondary electrons is much larger than the depth of maximum absorbed dose (Tubiana et al, 1956). Because of the differences in electron production and scattering, this modification depends on the nature of the plate, and larger modifications would be observed, if metallic plates were used.

In a similar way, with the Sagittaire linear accelerator the transition curves for 25-MV x-rays are modified by electrons set in motion in the lead collimator. The contribution of these electrons increases rapidly with field size and decreases with SSD (Figs. 5 and 6). It is responsible for the large variation of the depth of maximum absorbed dose which is observed (Table 1). This maximum depth is, for instance, 2.3 cm for a 30 cm × 30 cm field at 80 cm SSD, and 4.8 cm for a 5 cm × 5 cm field at 150 cm SSD (Marinello & Dutreix, 1973).

Similar variations have been observed with 6-MV x-rays: the depth of the maximum absorbed dose varies between 10 and 15 mm and the skin dose between 10 and 45%, depending on the experimental conditions (Aget, Doctoral thesis).

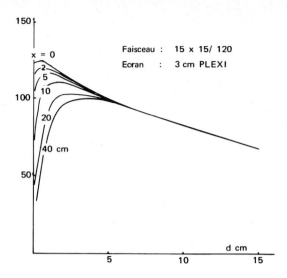

Figure 4. Percentage depth-dose curves in plexiglas for 24 MV x-rays. Field size, 15 cm X 15 cm; SSD, 90 cm. A 3-cm plexiglas plate was moved at distances ranging from 0 to 40 cm from the phantom surface. The curves are normalized at 10-cm depth.

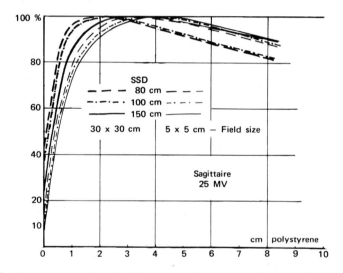

Figure 5. Build-up curves for 25 MV x-rays (Sagittaire) for two different field sizes (5 cm X 5 cm and 30 cm X 30 cm) and three different SSD (80, 100, and 150 cm).

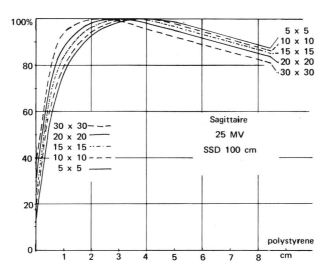

Figure 6. Build-up curves for 25 MV x-rays (Sagittaire), SSD=100 cm, for field sizes ranging between 5 cm X 5 cm and 30 cm X 30 cm.

TABLE 1. DEPTH OF THE MAXIMUM ABSORBED DOSE FOR DIFFERENT FIELD SIZES AND SSD WITH 25 MV X-RAYS (SAGITTAIRE).

Field Size (cm^2)	SSD (cm)		
	80	100	150
5	3.7	4.1	4.8
10	3.3	3.8	4.5
15	2.8	3.4	4.2
20	2.5	2.9	3.9
30	2.3	2.6	3.5

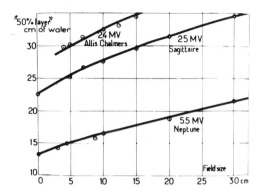

Figure 7. Variation of the "50% layer" in water as a function of field size for three different beams: linear accelerator Neptune, 5.5 MV x-rays; linear accelerator Sagittaire, 25 MV x-rays, and Allis-Chalmers betatron, 24 MV x-rays.

Such variation can be explained by considering that on a given point on the beam axis, the electron fluence in air depends on the solid angle of the collimator viewed from the point of measurement. This solid angle increases with field size and decreases with distance; furthermore, the secondary electrons produced in the collimator walls are emitted at various angles and their fluence decreases with distance due to their geometrical spread. The electron fluence cannot be reduced by adding a filter, which would stop the electrons from the collimator, but would emit an equivalent amount of electrons. The relative inefficiency of various metallic filters has been checked.

Variation of depth-dose curves. Scattering is certainly less important for high-energy x-rays than for cobalt γ-rays. However, its importance is generally underestimated, and it is even often assumed that the depth dose is independent of the field size (Cohen et al., 1972).

In order to estimate the effect of field size on depth-dose curves, we have measured the slope of the exponential part of depth-dose curves corrected for divergence, that is, for infinite SSD. The slope has been expressed as "50% layer," i.e., the thickness of water which reduces the dose to half its values; we have, in fact, measured the thickness which attenuates the depth dose from 80 to 40%. The results for two linear accelerators and one betatron are shown in Figure 7. For 25 MV x-rays, the 50% layer varies between 23 cm for "narrow beam" conditions to 34.5 cm for a 30 cm × 30 cm field; for 5.5 MV x-rays it varies between 13 and 21.5 cm.

Thus we would expect a larger depth dose when the field size is increased; however, in practice this variation is masked or even reversed as the depth of maximum dose becomes smaller. For depths of 5-10 cm the depth dose for large field sizes is surprisingly lower than for small field sizes. The interpretation of the depth-dose curves would be easier if the curves were normalized at a 10 cm depth.

SPECIFICATION OF BEAM QUALITY FOR RADIOTHERAPY

For high-energy x-rays, Podgorsak et al. (1975) have shown that the penetration of x-rays depends mainly on the nature and thickness of both target and flattening filters, the effect of the maximum energy being very slight above 20 MeV. The discrepancy in 50% layers between the 25 MeV linear accelerator and 24 MeV betatron (Fig. 7), is explained by such differences.

It is well established that an assessment of the beam quality of conventional x-rays requires not only the peak energy but the half-value layer in a reference medium; it is surprising to find out that for high-energy x-rays, one mentions only the maximum energy which is the most irrelevant parameter.

It has been proposed to specify high-energy beam quality by the depth of the 50% depth-dose for a given geometrical condition(SSD, 100 cm; field size, 10 cm X 10 cm). However, this depth depends not only on the beam attenuation, but also on the depth of maximum dose (as has been demonstrated), and it does not seem to be a convenient parameter for specifying the beam quality.

It is therefore suggested that the beam quality of high-energy x-ray beams be specified with a physical parameter actually related to attenuation. As narrow-beam conditions are difficult to achieve for high energies, the "50% layer" for a reference field size such as 10 cm X 10 cm could be a convenient parameter.

CLINICAL IMPLICATIONS OF PHYSICAL CHARACTERISTICS OF HIGH-ENERGY X-RAY BEAMS

Transition curves. The reaction of cutaneous and subcutaneous tissues is of interest in clinical applications, and the shapes of the transition curves have to be determined with accuracy for each type of acclerator.

When a deep tumor surrounded by healthy tissues has to be irradiated, the deeper the dose maximum, the better the dose distribution. However, when superficial nodes have to be included in the treated area, better dose distributions are achieved when the dose maximum is not too deep. For instance, mantle fields commonly used for the treatment of lymphoma include some relatively superficial lymph-node chains in the neck region. The 25-MV x-ray beam of the Sagittaire appears to be satisfactory for such treatments, since for a field larger than 30 cm X 30 cm, the dose maximum is not deeper than 2.6 cm, which means that the dose at a point 8-mm deep is as high as 90% of the maximum dose.

Dose variation at the beam edges. The beam edges represent a region of high-dose gradient where the dose distribution has to be determined carefully. When some critical organ appears to be close to the target volume, a sharp cut-off of the beam is important, as is generally the case for energies between 4 and 10 MV.

For higher energies, although the source sizes are usually rather small (a few millimeters in diameter), a broad penumbra region appears at the edge of the beam, even with a correctly aligned collimator. The observed dose variation is linked to the lack of lateral electronic equilibrium and can not be reduced by mechanical improvements. The width of the "penumbra" depends only on the x-ray's energy. Thus, for 6 MV the distance between 90% and 10% of the axis dose is as small as 5 mm, and this distance increases

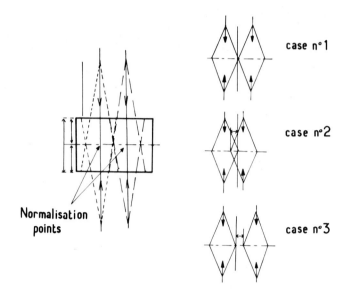

Figure 8. Two pairs of adjoining opposed beams are used with various distances between beam edges. (1) beams adjoining at the level of the patient midplane, (2) 1-cm overlapping of the beams, (3) beam edges separated by 1 cm.

with x-ray energy to about 20 mm for 20 MV x-rays (Bouhnik, Doctoral Thesis). Therefore the irradiation of small target volumes is better achieved with 4-10 MV x-ray beams than with very high-energy beams which give an undesirable dose around the target.

A sharp cut-off of the beam is especially inconvenient when adjoining fields are used, as illustrated in Figures 8 and 9. The dose along a line parallel to the beam axis at the level of the beam's junction has been computed in three cases: two adjoining beams, two overlapping beams and two separate beams. The percentage depth dose is normalized to 100 at the level of the patient midplane on the beam axis. Large under- and overdosages occur with small variations in the distances between beam edges, especially for 5.5 MV. In order to lessen the overdosage hazards, the position of the beam junction is usually moved with respect to the patient anatomy two or three times during the treatment course.

Obliquity and heterogeneity corrections. The magnitude of obliquity and heterogeneity corrections decreases when the beam energy increases (Dutreix, 1972); however, such corrections are still necessary for high-energy x-rays, when accurate dose distribution is needed. Thus a variation of 5 cm in tissue thickness leads to a dose variation as large as 20% for 5.5 MV and 14% for 25 MV.

Therefore, it is necessary to perform obliquity corrections even for high-energy x-ray beams. The so-called "isodose shift method" often used for cobalt-60 beams can be used for higher energy (Fig. 10). The isodose lines in the build-up region depend only on the tissue depth, and they remain parallel to the body contour whatever the obliquity. The deep isodose line has to be shifted by an amount $k.h$ where h represents the tissue deficit or the tissue excess and k, a factor depending on the beam energy. The k factor has been found

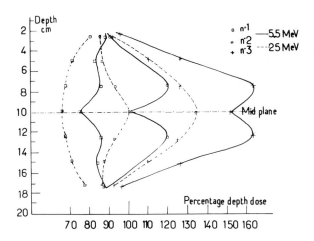

Figure 9. Percentage depth dose as a function of depth along a line parallel to the beam axis at the level of the beam junction for the three cases shown in Figure 8. Undotted lines: depth dose for 5.5 MV x-rays; dotted-lines: depth dose for 25 MV x-rays.

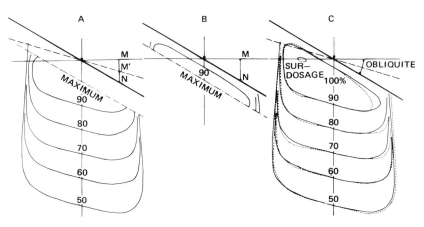

Figure 10. The isodose shift method applied to a 25-MV x-ray beam. (a) Deep isodose lines are shifted using a factor k=½. (b) Superficial isodose lines are shifted using a factor k=1 (they are parallel to the phantom contour). (c) Comparison of the shifted isodoses (undotted lines) with the measured isodoses (dotted lines).

to be equal to 2/3 for 5.5 MV x-rays as for cobalt (Dutreix & Dutreix, 1962) but for 25 MV x-rays, k=1/2 appears to be a better approximation (Marinello & Dutreix, 1973). These results are in good agreement with the calculations of Van der Giessen (1973).

In conclusion, uniform dose distributions can be more easily achieved with the use of high-energy x-ray beams and a greater precision can be expected. However, the difficulties encountered in high-energy beam dosimetry must not be underestimated. Accurate dose determinations are necessary for each type of accelerator for the full range of the geometric parameters used. Corrections to the basic dose distributions are necessary even with high energy, although the higher the energy, the smaller the correction.

REFERENCES

Cohen, M., D.E. Jones & D. Greene. 1972. Central axis depth dose data for use in radiotherapy. *Brit J Radiol* Supplement No. 11:8–17.

Dutreix, A. 1972. The influence of inhomogeneities and surface irregularities on accuracy. *In* Blickman, M.R. (ed) Proceedings of the Second Congress of the European Association of Radiology, Amsterdam, June, 1971. No. 249. Excerpta Medica, Amsterdam.

Dutreix, A. & J. Dutreix. 1962. Construction des isodoses pour les surfaces obliques et irrégulières. *J Radiol Electrol* 43:671–673.

Dutreix, J. & A. Dutreix. 1966. Etude comparée d'une série de chambres d'ionisation dans des faisceaux d'électrons de 20 et 10 MeV. *Biophysik* 3:249.

Dutreix, J., A. Dutreix & M. Tubiana. 1965. Electronic equilibrium and transition stages. *Phys Med Biol* 10:177–190.

Hettinger, G., C. Petterrson & H. Svensson. 1967. Displacement effect of thimble chambers exposed to a photon or electron beam from a betatron. *Acta Radiol TPB* 6:61–64.

Marinello, G. & A. Dutreix. 1973. Etude dosimétrique d'un faisceau de rayon X de 25 MeV. *J Radiol Electrol* 54:951.

Podgorsak, E.B., J.A. Rawlinson & H.E. Johns. 1975. X-ray depth doses from linear accelerators in the energy range from 10 to 32 MeV. *Amer J Roentgenol* 123:182–191.

Tubiana, M., J. Dutreix & A. Dutreix. 1956. Dispersion des electrons secondaires mis en mouvement par des rayons X de 22 MeV. *J Physique* 17:12A.

Van der Giessen, P.H. 1973. A method of calculating the isodose shift in correcting for oblique incidence in radiotherapy. *Brit J Radiol* 46:978–982.

High-Energy X-Ray Beams in the Management of Head and Neck and Pelvic Cancers

Carlos A. Perez, M.D.,
Professor of Radiology,
Division of Radiation Oncology,
Mallinckrodt Institute of Radiology,
Washington University School of Medicine,
St. Louis, Missouri

James A. Purdy, Ph.D.,
Instructor, Radiation Physics,
Division of Radiation Oncology,
Mallinckrodt Institute of Radiology,
Washington University School of Medicine,
St. Louis, Missouri

Alvin Korba, M.D.,
Fellow, Division of Radiation Oncology,
Mallinckrodt Institute of Radiology,
Washington University School of Medicine,
St. Louis, Missouri

William E. Powers, M.D.,
Professor and Director,
Division of Radiation Oncology,
Mallinckrodt Institute of Radiology,
Washington University School of Medicine,
St. Louis, Missouri

Since the early experience in clinical radiation therapy about 20 years ago with 1 MeV x-ray generators and cobalt-60 units, a great deal of reliance has been placed on the use of megavoltage beams for irradiation of cancer patients. A variety of machines have been manufactured, either in the form of betatrons or linear accelerators with energies ranging from 4 to 45 MV. In this paper we will analyze the physical characteristics and review the application of high-energy photons in the irradiation of patients with head and neck or pelvic malignancy. A great deal of the physical data was obtained by James Purdy, Ph.D., on an Allis-Chalmers betatron (23-MV x-rays), a Clinac-35 linear accelerator (25 MV x-rays) and a Clinac-4 accelerator (4 MV x-rays). The clinical data were derived from a retrospective review of selected groups of patients treated at the Mallinckrodt Institute of Radiology between 1954 and 1970, with histologically proven carcinoma of the nasopharynx, carcinoma of the tonsil, carcinoma of the uterine cervix, or adenocarcinoma of the prostate clinically localized to the pelvis. These patients were treated by 23 MV x-rays from an Allis-Chalmers betatron, alone or combined with 4 MeV x-ray or ^{60}Co beams, or in the case of carcinoma of the cervix, with intracavitary radioactive-source insertions (Ra-226, Co-60 or Cs-137).

PHYSICAL CHARACTERISTICS OF HIGH-ENERGY X-RAY BEAMS

Several physical factors affect the dose distribution of x- or γ-rays (Table 1).

The percentage depth-dose can be divided into various regions beginning with the surface dose and build-up region, the position of maximum dose, and the exponential portion of the depth-dose curve.

Recent measurements show that the surface dose for x-rays is essentially the same for energies ranging from 6 to 25 MV. These studies demonstrate that the dose distribution in the build-up region depends primarily on variables of the irradiation geometry, such as SSD, field size, and characteristics of secondary blocking trays. Figure 1 illustrates typical build-up curves for 4, 8 and 25 MV x-ray beams. The surface dose is 15-20% for a 10 X 10 cm field, increasing to 40-45% for a 35 X 35 cm field. Figure 2 illustrates the surface dose as a function of field size for a 25-MV x-ray linac beam.

TABLE 1. FACTORS AFFECTING DOSE DISTRIBUTION
OF X- OR γ-RAYS

1. Energy of beam (surface dose, depth dose characteristics)
2. Beam flatness (homogeneity-penumbra)
3. Depth of target volume
4. Tissue inhomogeneities
5. Irregular beam entrance surface
6. Accuracy of dose delivery (reproducibility of repeated setups)
   A. Patient immobilization
   B. Beam alignment
7. Beam Modifiers (wedges, blocks, compensators)

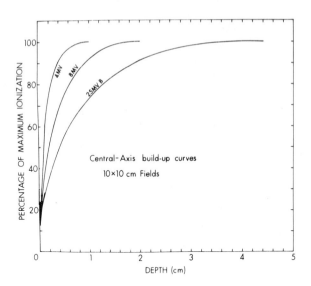

Figure 1. Examples of build-up curves for different x-ray beams.

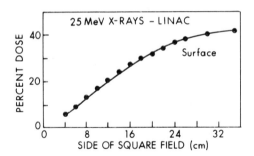

Figure 2. Surface dose as a function of field size. Note marked increase in surface dose as field size increases.

The depth of maximum dose varies from 1.5 cm for 6 MeV x-rays to about 4 cm for 25 MeV x-rays. However, measurements have shown that the position of the maximum dose cannot be unambiguously defined by the energy of the x-ray beams. Figure 3 illustrates the changing position of D_{max} as a function of field size for a 25-MV, linear-accelerator, x-ray beam. This shift of D_{max} is probably due either to low-energy photons or to electron contamination of the x-ray beam.

Beyond the build-up region, the percentage depth-dose varies considerably with photon energy. For example, for a 10 cm × 10 cm field, the central-axis depth-dose at 10-cm depth for a 6-MeV x-ray beam is 67%; for a 15-MV x-ray beam, 77%; for a 25-MV linac x-ray beam, 82%; and for a 25-MV betatron x-ray beam, 84%. An important point to emphasize is that the characterization of percentage depth-doses by beam energy is not

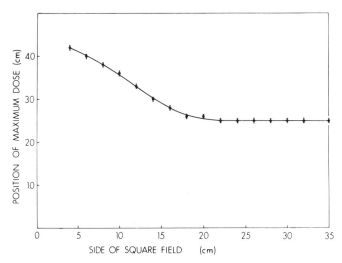

Figure 3. Position of maximum dose as a function of field size. Notice displacement toward the surface with larger fields.

adequate, and x-ray or electron beams should no longer be described strictly by energy. Instead, the depth-dose characteristics should be described by the surface dose, the position of the maximum dose depth, and the percentage depth-dose at 10 cm and 20 cm.

For x-ray beams from 6 to 45 MV, the angular distribution is forward and a flattening filter is required to produce a uniform dose at a specified depth. Outstanding field flatness over large fields (35 cm X 35 cm) can be obtained with linear accelerators, because the high output allows flexibility in the design of the flattening filter. On the Clinac-35 linear accelerator, a field flatness of ± 3% variation was achieved over 80% of a 28 cm X 28 cm field at a depth of 10 cm. Field symmetry, that is, the distribution of the dose around the central axis, is typically ± 2% for linear accelerators (Purdy, J.A., G.D. Oliver, Jr. and W.E. Powers, Unpublished Data).

The percentage depth-dose falls abruptly near the beam edge for high-energy x-rays due to the extremely small focal spot size (~ 3 mm) and because high-energy x-rays are scattered predominantly in the forward direction. It should be noted that the penumbra of a 25-MV x-ray beam is typically larger than that of a 6-MV x-ray beam. This will be further discussed in other chapters of these proceedings.

FACTORS AFFECTING TUMOR CONTROL BY IRRADIATION

Tumor control is a complex process influenced by a large number of factors, one of which is the physical distribution of the radiation in the target volume and surrounding sensitive structures. The probability of eradicating a tumor by irradiation can be affected by the following variables:

BIOLOGICAL FACTORS:

Tumor cell burden. Smaller tumors are more easily sterilized, and in general require slightly lower doses than larger lesions (Cohen, 1966).

Hypoxic cell subpopulation and re-oxygenation. Hypoxic cells are less sensitive to irradiation and appear in greater proportion in larger tumors. Fractionated irradiation induces regression of the tumor, increased vascularization and reoxygenation, with transfer of hypoxic cells into a well-oxygenated compartment (Kallman, 1972).

Marginal micro-extensions of the tumor. Malignant neoplasia invade the adjacent tissues by contiguity, along muscular or fascial planes or by permeation of lymphatics. These conditions demand irradiation of a larger area than would be required to encompass the bulk of the tumor alone.

The number of cells necessary to produce a tumor or a recurrence. It is not known exactly how many viable cells are necessary to produce a recurrence, but it is believed that a given course of radiation or cytotoxic agents does not necessarily destroy every cell in the tumor or the body. The host immune response is thought to play a role in the inactivation of small tumor-cell populations.

PHYSICAL FACTORS:

Dose-time-fractionation. Shukovsky and Fletcher (1973) showed that in carcinoma of the tonsil, there is a close correlation between the total dose of irradiation given and the percentage of local tumor control. Herring and Compton (1970), using data obtained by Shukovsky and Fletcher in Stage T_2 and T_3 carcinomas of the supraglottic larynx, postulated that relatively small variation in dose may significantly affect the probability of tumor cure. Fractionation of the irradiation and the time taken to deliver a number of fractions affect not only the rate of re-oxygenation of the tumor but also the recovery of normal tissues from irradiation (Cohen, 1966; Ellis et al., 1974; Wiernik, 1973).

Volume irradiated. The larger the volume treated, the smaller the total dose of irradiation that is tolerated by the normal tissues. Patterson (1963) correlated the incidence of skin necrosis with the dose of x-rays and the field size. Fletcher & Rutledge (1972) have noted poor tolerance and increased complications with extended periaortic ports for the treatment of carcinoma of the cervix.

Linear energy transfer (LET) and relative biological effect of various types of radiation. Doses needed to sterilize a tumor vary with the LET of the radiation used (Fowler, 1972).

HOST FACTORS:

Supportive systems. The tolerance of various tissues and organs in man to irradiation varies greatly. Circulating lymphocytes and gonadal germ cells are extremely sensitive, the epithelium of the digestive, respiratory and urinary tract is moderately sensitive, and some organs, such as muscle, long bones and central nervous system, tolerate higher doses of radiation without demonstrable deleterious effects (Cohen, 1966).

Host-tumor interrelationship. It is increasingly recognized that nutritional status and metabolism of the patient, the host immune response, and the organs involved by the neoplasia substantially affect the response of the tumor to a given dose of radiation.

ADVANTAGES OF HIGH-ENERGY PHOTON BEAMS

The physical characteristics of high-energy x-ray beams, and in the case of linear accelerators, the higher output, result in the following advantages in the treatment of deep-seated lesions:

1. Lower integral dose (better normal tissue tolerance),
2. Greater depth of maximum dose,
3. Better dose homogeneity in target volume, and
4. Less complicated delivery techniques for equal parameters (dose optimization).

Lower integral dose. According to Johns and Cunningham (1969), the integral dose delivered with betatron x-rays is about 80% of that for cobalt, regardless of the size of the patient and the field. With high-energy beams, there is no significant increase in integral dose as a function of the thickness of the patient or the field size.

Greater depth of maximum dose. Because of the depth of electronic build-up, there is significant sparing of the skin and superficial subcutaneous tissues.

Better dose homogeneity in target volumes. Due to the reduced effects of tissue heterogeneity and the greater depth-doses, a more homogenous dose can be delivered to the target volume, even with parallel, opposed portals, in a significant proportion of the patients.

Less complicated delivery techniques for equal parameters. Because opposing portals are adequate when high-energy photons are used, there is no need for complex treatment set-ups. This not only facilitates dose optimization in the tumor volume, but also allows for more accuracy in the repeated set-up of the patients treated with fractionated doses.

DISADVANTAGES OF HIGH-ENERGY PHOTON BEAMS

The disadvantage of high-energy photon beams include:

1. Greater depth of maximum dose (if not corrected when indicated),
2. High exit dose,
3. High initial cost,
4. High cost of supportive personnel, and
5. Expensive and relatively unavailable spare parts.

Greater depth of maximum dose. This is a double-edged sword, which can cause disastrous underdoses in patients with skin or subcutaneous involvement by tumor. Appropriate bolus must be used to bring the maximum dose to the surface, or low-energy beams should be combined with higher energies to eliminate this sparing effect of the superficial tissues.

High exit dose. It is not uncommon to see more severe skin reactions at the exit point than at the entrance port of a high-energy beam (Fig. 4).

High initial cost. Unless a large number of patients are anticipated, a high-energy photon machine may not be a sound economical investment. Equipment with lesser energies is a reasonable alternative, although as pointed out in this paper, it is theoretically not as desirable as higher energy machines.

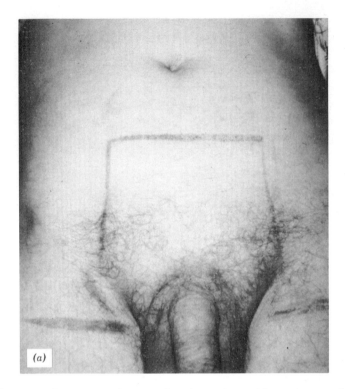

Figure 4. Frontal and posterior photographs of a patient with carcinoma of the prostate who received 6000 rads midplane dose in approximately 6½ weeks. Of the dose, 4000 rads were given through AP and PA ports, equal loading; the additional 2000 rads were delivered through AP ports only. Notice the significantly greater skin reaction in the buttock, which is due to the higher exit dose.

High cost of supportive personnel. In general, high-energy photon generators are complicated machines with easily disturbed electronic circuitry. These machines require constant attention from electronic technologists, electrical engineers and physicists.

Expensive and relatively unavailable spare parts. Because of the sparsity of the high-energy equipment, spare parts are generally expensive and sometimes difficult to obtain.

CONDITIONS FOR OPTIMAL EXTERNAL-BEAM IRRADIATION IN CLINICAL RADIATION THERAPY

As pointed out by Kitabatake et al. (1969), there are definite requirements for optimal dose distribution in both tumor and normal tissues, when using external irradiation. The following is a slightly modified list of factors published by these authors.

Small entrance and exit dose (except with superficial tumors). Ideally, when the maximum dose is not required at the skin or the subcutaneous tissues, the optimal dose

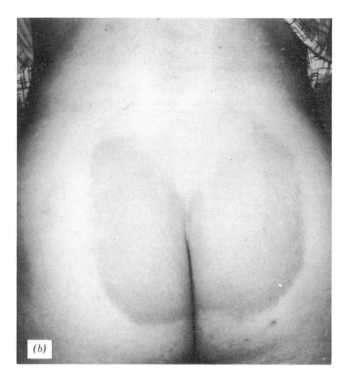

Fig. 4 (Continued)

distribution should be at the target volume in the depth of the patient, with lower dose to the skin, not only at the entrance, but also at the exit sites.

Small side-scattering dose. High-energy photon beams produce minimal amounts of side-scattered irradiation.

Small differential tissue absorption. It is known that with 250 kV x-rays there is a significantly greater absorption of irradiation in bone, compared with soft tissues. This phenomenon disappears with high-energy x-rays and increasing Compton effect between 1 and 10 MV. However, at energies of 20 MeV, a somewhat greater absorption of irradiation in bone is again observed (8-10%; Meredith, 1958).

Optimal tumor (target) dose. The aim of a good treatment plan is to exploit the maximum therapeutic ratio of a beam arrangement. The target volume should receive a homogenous dose while delivering as little dose as possible to the surrounding normal tissue.

Small integral dose. The ideal situation should be represented by an optimal dose to the target volume with a minimum dose contribution to the rest of the patient's normal tissues.

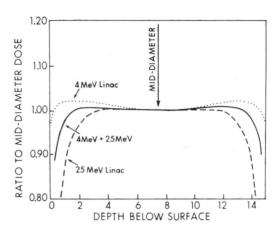

Figure 5. Dose profile with parallel, opposed ports in neck. With 4 MeV x-rays a high superficial dose will be required to achieve the desired midplane tumor dose. On the other hand, 25 MeV x-rays lend significant sparing of the skin and subcutaneous tissues, but could appreciably underdose lymph node metastases. A satisfactory dose distribtution is achieved by a combination of 4 MeV and 25 MeV x-rays solid time.

CLINICAL EXPERIENCE IN THE TREATMENT OF HEAD AND NECK CANCER

Relatively similar techniques have been employed since 1959 for the treatment of patients with carcinoma of the nasopharynx or the tonsil. The treatment consisted of delivering about 3000 rads T.D. with cobalt-60 or 4 MeV x-rays to large, lateral, opposing portals that included the primary lesions and all of the neck lymph nodes. Following this, an additional 3000-3500 rads T.D. were given with 23-MV betatron x-rays through smaller, lateral, opposing ports that covered the primary lesions and the upper-neck lymph nodes. In addition, the lower neck received 2000-3000 rads calculated at 3 or 4 cm through an AP tangential port with cobalt-60 or 4-MeV x-rays for a total of 5000-6000 rads. The daily tumor dose was 180-200 rads, five fractions given per week (Perez et al. 1969, 1972).

Figure 5 shows the advantage of combining low and high-energy x-ray beams for the treatment of the primary tumor and upper neck. If cobalt-60 or 4-MeV x-rays were used to deliver the entire dose, the temporo-mandibular joints and the mandible would receive doses in excess of 7000 rads, which would lead to severe complications. On the other hand, using 23-MeV x-rays only would cause a substantial underdose to the neck lymph nodes, which is even more critical if they are fixed to the skin and subcutaneous tissues.

Figures 6a, b and c demonstrate the typical isodose distribution for 4-MeV x-rays, 22 MeV x-rays or a combination of both. In patients with unilateral lesions without contralateral lymph nodes, the contribution from the 23-MeV x-ray beam can be given from one side only, decreasing the dose of radiation given to the contralateral mandible and parotid gland. In patients with extensive tumors or large, fixed, lymph nodes, the addition of an electron beam "boost" further optimizes the dose distribution (Fig. 7). In small T_1 tonsillar lesions, adequate treatment can be delivered with ipsilateral x-rays, electron beams or a pair of wedges. This spares the contralateral normal tissues.

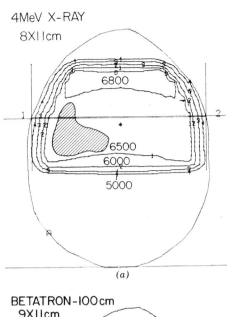

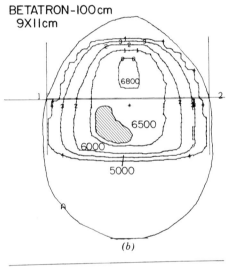

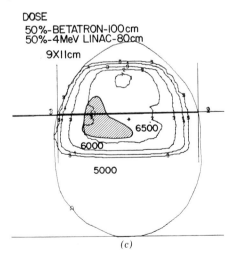

Figure 6. Examples of treatment plans for a carcinoma of the tonsil using (a) 4 MeV x-rays (b) 22 MeV x-rays or (c) a combination of both.

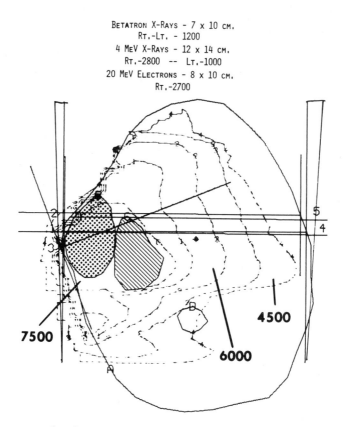

Figure 7. Treatment plan for a patient with a carcinoma of the tonsil and a large lymph node in the ipsilateral upper neck. An adequate tumor dose is obtained with a combination of 4 MeV and 23 MeV x-rays, and an electron beam boost (20 MeV) to the lymph node and the primary. Contralateral lymph nodes are electively irradiated and receive 4500 rads. The wedges are used to reduce the dose to the adjacent oral cavity.

Table 2 summarizes the three-year, absolute, survival rate and incidence of recurrence in a group of 79 patients treated for nasopharyngeal carcinoma. It is obvious that the doses of radiation given were not adequate to control lesions more extensive than $T_{1,2}$ or $N_{1,2}$. This may be due in part to the doses of radiation given, usually in the range of 6000 rads. Table 3 summarizes the recurrences in the necks of these patients. The tumor control in the lymph nodes was excellent (less than 20% failure rates), except for large fixed nodes (40% recurrence rates).

Major complications after irradiation have been relatively few (Table 4). This may be related to the lower dose of radiation given, but more likely, to the use of high-energy beams to deliver a portion of the dose.

Table 5 summarizes the three-year, absolute, survival rates for patients with carcinoma of the tonsil. Three-year survival rates of 76% were noted in cases of $T_{1,2}$–N_0 tumor and approximately 45% for patients with $T_{1,2,3}$–$N_{0,1,2}$ lesions. Lower survival (10-20%) was

TABLE 2. CARCINOMA OF THE NASOPHARYNX: THREE-YEAR ABSOLUTE SURVIVAL WITH NO EVIDENCE OF DISEASE[a]

Stages	Survival	Recurrences			Positive Neck Nodes Only	DM	ICD[c]
		Local or Marginal					
		Alone	Positive Neck Nodes	Positive DM^b			
$T_{1,2}-N_0$	12/17 (70%)	2	1	3	1	3	2
$T_{1,2}-N_{1,2}$	13/27 (48%)	3	2	4	2	3	
$T_{1,2}-N_3$	1/10 (10%)	1	2	3	1	2	
$T_3-N_{0,1,2}$	3/14 (21%)	5	1	2	1	1	2
T_3-N_3	1/5 (20%)	1	1	1	1		
T_4, Any N	2/11 (22%)	5	1	3		1	

[a]Mallinckrodt Institute of Radiology, 1951-1972
[b]DM=Distant Metastases
[c]Intercurrent Disease

noted in the more extensive lesions ($T_{3,4}$). Furthermore, the local failure rate was approximately 20% in the $T_{1,2}$ tumors, about 30% in T_3 lesions, and 70% in T_4. As in the nasopharyngeal tumors, the percent of neck node recurrence was below 20% for $N_{0,1}$ lymph nodes and about 30% for $N_{2,3}$ lymph nodes. With doses of 5000 rads, only one of 97 patients developed a contralateral neck recurrence (Table 6).

The control of the primary tumor in these patients is comparable to other series (Scanlon et al., 1967a,b; Wang & Meyer, 1971; Wang, 1972), but not as high as reported by some authors (Shukovsky & Fletcher, 1973). This is probably related to the moderate doses of radiation used in the earlier period of treatment and should not be construed as a lack of effectiveness of high-energy x-ray beams. Shukovsky and Fletcher (1973) have shown that in the tonsil, more extensive lesions (T_3 and T_4) require doses in excess of 7000 rads for a high probability of local control. However, as shown by Gelinas and

TABLE 3. CARCINOMA OF THE NASOPHARYNX: RECURRENCE IN IRRADIATED NECKS[a]

Staging	Number of Patients	Ipsilateral	Contralateral	Bilateral	(Percent)
N_0	20	1		1^b	10%
N_1	16	1	1	1^b	18%
N_2	28	3	1		14%
N_3	15	5		1	40%

[a]Mallinckrodt Institute of Radiology, 1951-1972; Unlimited follow-up, minimum 3 years
[b]6 X 6 cm upper ports, 250 Kv (1955)

TABLE 4. CARCINOMA OF THE NASOPHARYNX: MAJOR COMPLICATIONS AFTER IRRADIATION[a]

Osteonecrosis		
Nasal bones	1	
Maxillary antrum	1	250 kv
Zygoma	1	
Mandible	2	
Carotid Rupture	2[b]	
Pharyngo-cutaneous fistula and carotid rupture	1[b]	
T–M joint ankylosis	3	
Marked neck fibrosis	2	

[a] Mallinckrodt Institute of Radiology, 1951-1973
[b] Surgery for suspected neck recurrence (only one patient showed tumor in nodes, not viable)

Fletcher (1973) in a retrospective analysis of failures after irradiation of 125 patients with carcinoma of the tonsil, in five cases the recurrence was due to a geographical miss and in seven to underdose, but in 11 cases no obvious reason could be identified. It should be noted that the complications in the patients reported by us (Table 7) were significantly less than those described by other authors using cobalt-60 or 2-MeV x-rays (Cheng & Wang, 1974; Grant & Fletcher, 1966).

TABLE 5. CARCINOMA OF THE TONSIL TREATED BY RADIATION ALONE: THREE-YEAR ABSOLUTE SURVIVAL, NO EVIDENCE OF DISEASE[a]

Stages	Survival	Recurrences Local or Marginal Alone	Positive Neck Nodes	Positive DM[b]	Neck Nodes Only	DM	ICD[c]
$T_{1,2}$–N_0	16/21 (76%)	3	1		1		2
$T_{1,2}$–$N_{1,2}$	4/9 (44.4%)	1		2	1	2	
T_3–$N_{0,1,2}$	10/22 (45.5%)	4		1	2	2	2
$T_{1,2,3}$–N_3	4/16 (25%)	1	4		2	1	2
T_4–Any N	3/27 (11%)	9	5	3	2	2	3

[a] Mallinckrodt Institute of Radiology, 1954-1972 (3 patients salvaged by operation are included and 3 who died during treatment are tabulated as failures)
[b] DM=Distant Metastases
[c] Intercurrent Disease

TABLE 6. CARCINOMA OF THE TONSIL TREATED BY IRRADIATION ALONE: CLINICAL LYMPH-NODE STAGING AND PERCENTAGE OF NODAL RECURRENCE

Staging	Number of Patients	Recurrences Ipsilateral	Contralateral
N_0	33	2 (6%)	
N_1	21	5 (23%)	
N_2	10	3 (30%)	
N_3	33	10 (30%)	1

TABLE 7. CARCINOMA OF THE TONSIL TREATED BY RADIATION THERAPY ALONE: COMPLICATIONS (UNLIMITED TIME)

Fatal	
Carotid rupture[a]	1
Post-operative cerebrovascular accident[b]	1
Died during treatment, undetermined cause	2
Non-fatal	
Severe xerostomia	7
Severe dental decay	4
Osteonecrosis of mandible	3
Fistula[b]	3
Difficulty swallowing	3
Edema of larynx	2
Total	97

[a]Massive tumor
[b]Operation for suspected persistent tumor

Some therapists are concerned with the build-up of high-energy beams and point out that the superficial tissues are receiving lower doses. However, as noted in the chapter, the dose at 1 cm with 23-MeV x-rays is over 90%. In point of fact, the control of the lymph nodes in the neck has been high, even though they were relatively superficial. It is important to recognize that if the lymph nodes in the neck are fixed to the skin, bolus must be used, at least for a portion of the treatment, in order to enhance the surface dose.

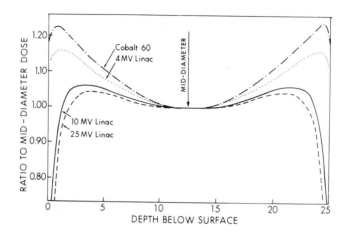

Figure 8. Dose profile for parallel, opposing pelvic ports with different energies. Note that significantly greater doses must be delivered with cobalt 60 and 4-MeV x-rays, as compared to higher energy, x-ray beams.

EXPERIENCE IN THE TREATMENT OF CARCINOMA OF THE CERVIX

The high propensity of carcinoma of the cervix to metastasize to the pelvic lymph nodes is well known (Fletcher, 1970). Because of this, in addition to intracavitary radium, which in most instances can adequately treat the primary tumor, external-beam irradiation has been given to the pelvis, to deliver cancerocidal doses to the lymph nodes.

Because of the thickness of the pelvis, high-energy beams are specially suited for this treatment. Figure 8 shows the dose profile for parallel ports, anterior/posterior and posterior/anterior, pelvic beams using different x-ray beam energies. It is evident that with lower-energy beams, higher doses must be given, and more complicated field arrangements used for achievement of the same midplane tumor dose. This, by necessity, delivers more irradiation to the bladder and the rectum, which may result in a higher complication rate. Beams above 20 MeV can provide adequate dose distribution with opposing anterior/posterior and posterior/anterior ports, and there is no need to use lateral ports, except in obese patients, thus decreasing the amount of small bowel that is irradiated. Even using the four-port "box" technique, the dose distribution with 22-MeV x-rays is superior to that with 6-MeV x-rays (Figs. 9a and b). Occasionally, a three-field arrangement with an AP and two posterior oblique portals, produces excellent dose distribution (Figs. 10a and b), albeit the high-energy beams are superior to 6-MeV x-rays.

At our institution we employ a modification of the Manchester technique to treat carcinoma of the cervix. The following doses have been used on most patients.

For Stage I or IIA, 1000 or 2000 rads are given to the whole pelvis through AP and PA, 15 cm X 15 cm ports. In addition, 3000-4000 rads are given to the parametria through similar ports with a step-wedge midline block. For Stages IIB and III, 2000 rads are given to the whole pelvis, followed by 4000 rads split fields with midline step-wedge block.

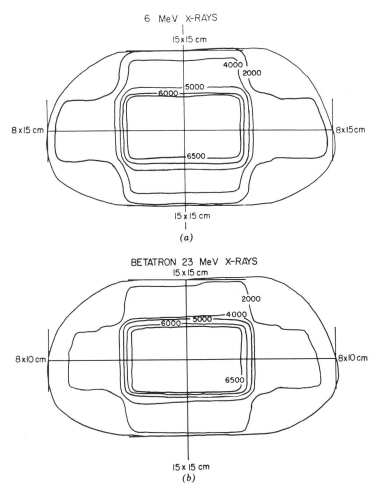

Figure 9. Isodose curves for the four-field technique using: (a) 6-MeV x-rays or (b) 23-MeV x-rays. Note that in order to deliver the same dose to the volume of interest (6500 rads), a significantly larger volume of tissue must receive at least 4000 rads when 6-MeV x-rays are used.

External beam irradiation is combined with 7500-8000 mg hours (about 7500 rads to point A) in two intracavitary applications. This technique allows the delivery of a high central pelvic dose and a homogenous dose of approximately 6000-7000 rads throughout the entire pelvis (Fig. 11). With the midline step wedge, the amount of irradiation delivered to the bladder and the rectum is minimized, without compromising the parametrial dose (Fig. 12).

Table 8 shows the five-year, disease-free survival for about 200 patients with Stages I through III carcinoma of the cervix. In Stages I and IIA, the results are comparable to those reported by other authors (Kline et al., 1969; Selim et al., 1974). In Stages IIB and III, the survival rates are comparable to the best reported by other groups using high-

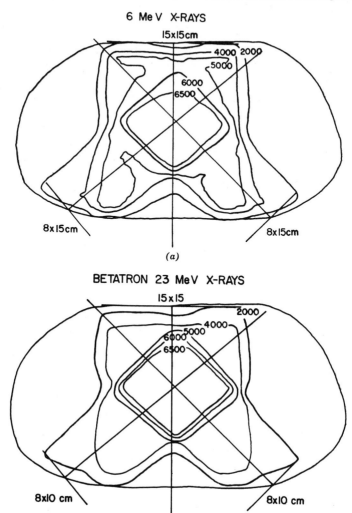

Figure 10. Treatment plans for three-field technique with: (a) 6-MeV x-rays or (b) 23-MeV x-rays. Again, a larger volume of the pelvis must receive at least 5000 rads in order to deliver 6500 rads to the midplane of the pelvis when using 6-MeV x-rays. The same volume receives only about 4000 rads when using 23 MeV x-rays, even though the same midplane dose distribution is accomplished.

energy beams (Fletcher, 1971) and superior to those reported in several publications (Arneson & Williams, 1960; Kline et al., 1969; Marcial & Bosch, 1968; Selim et al., 1974). The pelvic failure rate is particularly low in patients with Stages IIB and III tumors, which is proof of the effectiveness of high-energy beams in controlling pelvic tumor.

Table 9 shows the complications of treatment. Before 1963, a higher incidence of major complications was noted in a group of patients treated for Stages IIB and III carcinoma of the cervix, because of technical difficulties integrating the intracavitary therapy

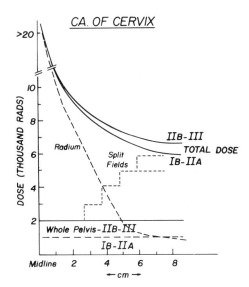

Figure 11. Dose profile for the Mallinckrodt Institute of Radiology technique for the treatment of carcinoma of the cervix using a combination of intracavitary radioactive-source applications and external beam, 23-MeV x-rays. A dose in excess of 9000 rads is delivered through the midline (within 2 cm from the sources), and a homogenous 6000-7000 rads is given to the entire parametrial tissues.

and the external beam due to an inadequate design of the wedge filter which resulted in excessive irradiation to the bladder and the rectum. However, after this was corrected, the complication rates were in the range of 6-7% for the early stages and 11% for more advanced stages. With orthovoltage or even cobalt-60 beams, pathological fractures of the femoral neck or necrosis of the femoral head have been reported (Mercado & Sala, 1968).

The superiority of high-energy beams in the treatment of carcinoma of the cervix was strongly suggested by Allt (1969), comparing cobalt-60 and betatron x-rays in a randomized group of patients with Stages IIB and III carcinoma of the cervix. The incidence of complications was significantly lower in the group of patients treated with betatron x-rays.

CARCINOMA OF THE PROSTATE

Since 1967 approximately 125 patients have been treated with curative intent for inoperable carcinoma of the prostate localized to the pelvis. Patients are carefully screened for tumor dissemination with alkaline and acid phosphatase determinations, bone scan and pedal lymphangiogram (Perez et al., 1974). The basic treatment consists of 5000 rads midplane tumor dose to the common iliac, external iliac and hypogastric lymph nodes through anterior/posterior and posterior/anterior 13 cm X 15 cm portals. An additional 1000 rads are delivered to the external iliac and hypogastric nodes with a 14 cm X 14 cm anterior-posterior portal, and an additional 1000 rads to the prostate only with a 6 cm

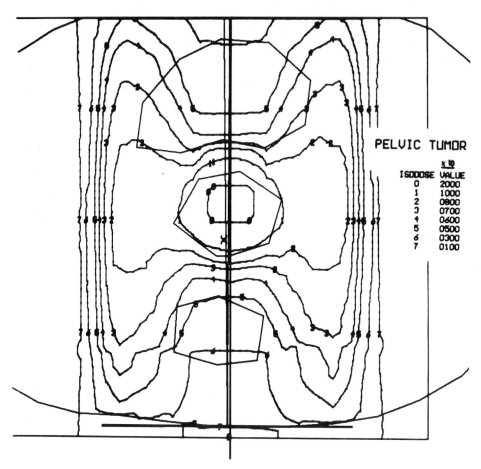

Figure 12. Isodose curves for the treatment technique shown in Figure 11 demonstrate the relative sparing of the bladder and the rectum with the use of a midline shield after 1000 or 2000 rad whole-pelvis irradiation. A step-wedge allows an integration of the intracavitary and external beam dose with optimal irradiation of the parametria.

X 8 cm or 8 cm X 10 cm anterior-posterior portal. The dose fractionation is 180 rads TD daily, 5 fractions per week. In approximately 20% of the patients who have received 5000 rads TD, more severe intestinal and urinary side-effects of treatment make it necessary to interrupt therapy for one or two weeks, before it can be completed. Figures 13a and b show the dose distribution to the midplane of the pelvis and a dose profile comparison for 25-MeV, linear-accelerator x-rays and 23-MeV, betatron x-rays. Because of the better depth-dose obtained with the betatron, this modality has been preferred.

Tables 10 and 11 show the cumulative survival rate and site of failure in 86 patients with Stage C and 12 patients with Stage B tumors. Some of these patients had been diagnosed several years prior to the radiation therapy, and about 50% had been previously treated with orchiectomy and estrogen therapy. The cumulative five-year survival rate is slightly over 40% for Stage C and 80% for Stage B patients. This compares favorably with the series reported by Ray et al., (1973), delivering 7000 rads with the 6-MeV linear

TABLE 8. RADIATION THERAPY OF CARCINOMA OF THE UTERINE CERVIX: PRELIMINARY (PARTIAL) RESULTS[a]

Clinical Stage	Number of Patients	Alive, NED[b] 5 Years	Pelvic Recurrence Alone	Pelvic Recurrence + DM[c]	% Pelvic Failure	DM Only
I	88	73 (83%)	1	5	6.9	1
IIA	28	17 (61%)		3	10.7	2
IIB	73	45 (61%)	4	5	12.3	3
III	64	32 (50%)	7	8	23.4	5

[a] Mallinckrodt Institute of Radiology [b] NED=No evidence of disease
[c] DM=Distant Metastases

TABLE 9. CARCINOMA OF THE CERVIX: MAJOR COMPLICATIONS

	Stage I	Stage IIA	Stage IIB Before 1963	Stage IIB After 1964	Stage III Before 1963	Stage III After 1964
Number of cases	86	28	36	37	37	27
Number of complications	5	2	10	4	11	3
Percent	5.8%	7%	27%	11%	29%	11%
Complications						
G.I.						
Recto-vaginal fistula	1[a]		2	1	4	1
Recto-uterine fistula			1			
Sigmoid perforation/ stricture		1		1[b]		
Rectal ulcer/proctitis	1	1	3	1	3	1
Intestinal obstruction or perforation	2				3	1
G.U.						
Vesico-vaginal fistula			1[c]			
Ureteral obstruction	1		1			
Severe cystitis			2		1	
Other (Not directly related to irradiation)						
Hepatic necrosis				1		
Pulmonary embolism			1			
Thrombophlebitis	1 (Non-fatal)					

[a] Also ureteral stricture (following post-irradiation hysterectomy for pelvic abscess)
[b] Also ureteral stricture and small bowel obstruction 2 months later
[c] Concomitant recto-vaginal fistula

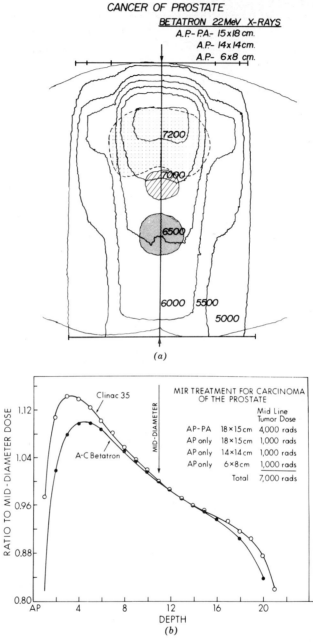

Figure 13. Isodose distribution (a) and dose profile (b) for a patient with adenocarcinoma of the prostate treated with progressively decreasing fields using 22 MeV x-rays.

accelerator. Table 12 shows the percentage of pelvic failure, approximately 20% in the patients with Stage C treated before December, 1973. Only three of the patients with Stage B have failed, two of them in the prostate. The majority of the patients failing

TABLE 10. CARCINOMA OF THE PROSTATE, STAGE C: ABSOLUTE SURVIVAL RATE, WITH NO EVIDENCE OF DISEASE[a]

	Years After Treatment				
	1	2	3	4	5 or more
Total number patients	83	67	54	42	34
Cumulative survival without clinical tumor	75 (90%)	47 (70%)	27 (50%)	18 (43%)	14 (41%)
Dead with tumor[b]	7	12	6	1	2
Dead of intercurrent disease	1	0	1	1	1
Dead of second primary	0	0	0	1	2

[a]Mallinckrodt Institute of Radiology, 1966-1973
[b]Includes three patients alive with tumor

TABLE 11. CARCINOMA OF THE PROSTATE, STAGE B: ABSOLUTE SURVIVAL, NO EVIDENCE OF DISEASE[a]

	Years after treatment				
	1	2	3	4	5
Number of patients	12	7	5	3	1
Deaths:					
Local Recurrence		1	1		
Distant Metastasis		1			
Intercurrent disease				1	

[a]Mallinckrodt Institute of Radiology, 1966-1973

TABLE 12. CARCINOMA OF THE PROSTATE, STAGE C: SITE OF FAILURE AFTER TREATMENT[a]

	Years at Risk				
	1	2	3	4	5
Total number patients	83	67	54	42	34
Percent all local recurrences	19.3%	18%	11%	4.7%	3%
Local recurrence only	2	3	2	1	
Local recurrence and distant metastasis	2	3	2		1
Distant metastasis only	3	4	3		1

[a]Mallinckrodt Institute of Radiology, 1966-1973
(One patient died during treatment of congenital heart failure. During the first year, only progression of tumor was considered failure)

TABLE 13. ADENOCARCINOMA OF THE PROSTATE, RADICAL IRRADIATION: COMPLICATIONS IN 95 PATIENTS[a]

Minor	
Persistent diarrhea or rectal symptoms	5
Chronic symptoms of cystitis	3
Urinary incontinence	6[b]
Pubic subcutaneous fibrosis	3
Major	
Pubic bone necrosis	1
Proctitis treated with colostomy[c]	1
Peri-rectal abscess (4 years, after treatment)	1
Impotence	8

[a]Mallinckrodt Institute of Radiology, 1967-1973
[b]Patients had TUR's
[c]Hemorrhoidectomy performed while patient was receiving radiation therapy

developed distant metastases. Table 13 lists the complication rate on the 95 patients reported. There is an extremely low incidence of severe complications, despite the high dose of irradiation delivered. Other treatment techniques used for prostatic irradiation consist of rotational therapy with linear accelerators (Ray et al., 1973), opposing pelvic portals with an appositional perineal port (del Regato, 1967) or three-field techniques. The dose distribution with high-energy beams is superior, since the dose delivered to the rectum is significantly less.

CONCLUSIONS

It is obvious that for the achievement of maximum dose optimization in radiation therapy, several beams of different energies need to be combined. It is only in very superficial or deep-seated lesions that a single-energy beam may yield the best results. Thus, precise treatment-planning and dosimetry are vital to optimal radiation therapy, as are adequate evaluation of the patient, accurate tumor-localization techniques, and precise and reproducible patient repositioning and treatment delivery.

High-energy beams have distinct advantages for the treatment of deep-seated lesions in the head and neck, thorax and pelvis. They allow the delivery of homogenous high doses to the tumor, with relative sparing of the normal tissues. In cases of superficial lesions, combination with lower-energy beams provides the best dose distribution. High-energy beams have the disadvantage of skin-sparing effects and lower doses in the superficial subcutaneous tissues. Thus, when the skin and the subcutaneous tissues are involved, the use of bolus is mandatory to avoid underdosing. The addition of electron beams adds greater capability to the radiation therapist's ability to achieve optimal tumor doses.

Linear accelerators with high-energy beams have particular applications in the treatment of patients requiring extended-field techniques, such as in carcinoma of the cervix with periaortic metastatic nodes, carcinoma of the prostate, or testicular tumors. The high exit dose may sometimes prove to be a technical difficulty, because more marked skin reactions may be observed at the exit rather than the entrance site.

REFERENCES

Allt, W.E. 1969. Supervoltage radiation treatment in advanced cancer of the uterine cervix: A preliminary report. *Canad Med Assoc J* 100:792–797.

Arneson, A.N. & C.F. Williams. 1960. Long-term follow-up observations in cervical cancer. *Amer J Obst Gynecol* 80:775–790.

Chen, K.Y. & G. H. Fletcher. 1974. Malignant tumors of the nasopharynx. *Radiology* 99:165–171.

Cheng, V.S. & C.C. Wang. 1974. Osteoradionecrosis of the mandible resulting from external megavoltage radiation therapy. *Radiology* 112:685–689.

Cohen, L. 1966. Radiation response and recovery: Radiobiological principles and their relation to clinical practice. p. 208–348. In Schwartz, E.E. (ed.) The Biological Basis of Radiation Therapy. J.B. Lippincott Co., Philadelphia and Toronto.

del Regato, J.A. 1967. Radiotherapy in the conservative treatment of operable and locally inoperable carcinoma of the prostate. *Radiology* 88:761–766.

Ellis, F., A. Sorenson & C. Lescrenier. 1974. Radiation therapy schedules for opposing parallel fields and their biological effects. *Radiology* 111:701–707.

Fletcher, G.H. 1971. Cancer of the uterine cervix: Janeway lecture, 1970. *Amer J Roentgenol* 111:225–242.

Fletcher, G.H. 1972. Elective irradiation of subclinical disease in cancers of the head and neck. *Cancer* 29:1450–1454.

Fletcher, G.H. & F.N. Rutledge. 1972. Extended field technique in the management of the cancers of the uterine cervix. *Amer J Roentgenol* 114:116–122.

Fowler, J J. 1973. Review of the basis of heavy particle therapy and past clinical trials. p. 28–38. In Proceedings of the Conference on Particle Accelerators in Radiation Therapy, Los Alamos, New Mexico, 1972. National Technical Information Service, Springfield, Va.

Gelinas, M. & G.H. Fletcher. 1973. Incidence and causes of local failure of irradiation in squamous cell carcinoma of the faucial arch, tonsillar fossa and base of the tongue. *Radiology* 108:383–387.

Grant, B.P. & G.H. Fletcher. 1966. Analysis of complications following megavoltage therapy for squamous cell carcinomas of the tonsillar area. *Amer J Roentgenol* 96: 28–36.

Herring, D. F. & D.M. Compton. 1970. The degree of precision required in the radiation dose delivered in cancer radiotherapy. *Enviromed Report, No. 216.*

Kallman, R.F. The phenomenon of reoxygenation and its implications for fractionated radiotherapy. *Radiology* 105:135–142.

Johns, H. & J. Cunningham. 1969. The Physics of Radiology. 3rd Edition. Edited by Milton Friedman. Charles C. Thomas Co., Springfield, Ill.

Kitabatake, T., H. Hattori & Y. Okumura. 1969. Optimum energy in supervoltage x-ray therapy. *Strahlentherapie* 137:158–161.

Kline, J.C., A.E. Schultz, H. Vermund & B.M. Peckham. 1969. High-dose radiotherapy for carcinoma of the cervix: Methods and results. *Amer J Obst & Gynecol* 104:479–484.

Levitt, S.H. 1973. Characteristics of Megavoltage External Beams. University of Minnesota Press, Minneapolis.

Marcial, V.A. & A. Bosch. 1968. Fractionation in radiation therapy of carcinoma of the uterine cervix: Results of prospective study of 3 vs 5 fractions per week. p. 238–249. *In* Vaeth, J.M. (ed.) Frontiers of Radiation Therapy and Oncology. Vol. 3. S. Karger, Basel, New York.

Mercado, R. Jr. & J.M. Sala. 1968. Comparisons of conventional and supervoltage radiation in the management of cancer of the cervix: Analysis of survival rates and complications. *Radiology* 90:867–970.

Meredith, W.J. 1958. Some aspects of supervoltage radiation therapy. *Amer J Roentgenol* 79:57–63.

Moench, H.C. & T.F. Phillips. 1972. Carcinoma of the nasopharynx: Review of 146 patients with emphasis on radiation dose and time factors. *Amer J Surg* 124:515–518.

Patterson, R. 1963. The Treatment of Malignant Disease by Radiotherapy. The Williams and Wilkins Co., Baltimore.

Perez, C.A., L.V. Ackerman, W.B. Mill, J.H. Ogura & W.E. Powers. 1969. Cancer of the nasopharynx: Factors influencing prognosis. *Cancer* 24:1–17.

Perez, C.A., L.V. Ackerman, W.B. Mill, J.H. Ogura & W.E. Powers. 1972. Malignant tumors of the tonsil: Analysis of failures and factors affecting prognosis. *Amer J Roentgenol* 114:43–58.

Perez, C.A., L.V. Ackerman, I. Silber & R.K. Royce. 1974. Radiation therapy in the treatment of localized carcinoma of the prostate. Preliminary report using 22 MeV photons. *Cancer* 34:1059–1068.

Ray, G.R., J.R. Cassady & M.A. Bagshaw. 1973. Definitive radiation therapy in carcinoma of the prostate. *Radiology* 106:407–418.

Scanlon, P.W., R.E. Rhodes Jr., L.B. Woolner, K.D. Devine & J.B. McBean. 1967. Cancer of the nasopharynx: 142 patients treated in the 11 year period 1950–1960. *Amer J Roentgenol* 99:313–325.

Scanlon, P.W., K.D. Devine, L.B. Woolner & J.B. McBean. 1967. Cancer of the tonsil: 131 patients treated in the 11 year period 1950 through 1960. *Amer J Roentgenol* 100:894–903.

Selim, M.A. J.L. So-Bosita, A.B. Little & W. Topolnicki. 1974. Carcinoma of the cervix: Clinical experience during a 10-year period, 1958–1967. *Ob/Gyn* 44:77–83.

Shukovsky, L.F. & G.H. Fletcher. 1973. Time-dose and tumor volume relationships in the irradiation of squamous cell carcinoma of the tonsillar fossa. *Radiology* 1973: 621–626.

Wang, C.C. 1972. Management and prognosis of squamous cell carcinoma of the tonsillar region. *Radiology* 104:667–671.

Wang, C.C. & J.E. Meyer. 1971. Radiotherapeutic management of carcinoma of the nasopharynx: An analysis of 170 patients. *Cancer* 28:566–570.

Wiernik, G. 1973. The significance of the time-dose relationship in radiotherapy. *Brit Med Bull* 29:39–43.

Electron-Beam Dose Distribution in Inhomogeneous Media

W. Pohlit, Ph.D.,
Max Planck Institute of Biophysics,
University of Frankfurt/Main,
Frankfurt, Germany

K. H. Manegold, Ph.D.,
Max Planck Institute of Biophysics,
University of Frankfurt-Main,
Frankfurt, Germany

INTRODUCTION

The success of radiation therapy depends on the homogeneity of the dose distribution in the irradiated tumor and on sufficient reduction of absorbed dose in critical organs located in the irradiated area. Especially in electron treatments, inhomogeneous dose distributions occur and have to be evaluated quantitatively so that optimal irradiation conditions can be obtained.

These inhomogeneities in the distribution of absorbed dose in an irradiated patient are due mainly to variations in the density of the material traversed by the fast electrons. These effects can be taken into consideration by a simple method which has been recommended by the International Commission on Radiation Units (ICRU), as long as the inhomogeneous material is present in the form of a uniform large slab (see below). This, however, is not usually the case. Therefore in this discussion a quantitative approach is given for the management of more complicated inhomogeneities, where electron scattering is involved. Under these conditions a relatively simple method for determining absorbed dose distribution can be developed, and used in computer programs. In this way, an obscure difficulty in electron therapy, the occurrence of so-called "hot spots" and "cold spots" in absorbed dose distribution, can be handled with sufficient accuracy to improve the overall value of electron therapy.

ABSORBED DOSE DISTRIBUTION BEHIND UNIFORM SLABS

For large and uniform slabs, corrections for the change of the depth-dose curve and of the absorbed dose distribution have been well known for several years (Pohlit, 1960, 1969; Laughlin, 1965; Laughlin et al., 1965; Boone et al., 1969; Dahler et al., 1969). As recommended also by the ICRU (1972) a coefficient of equivalent thickness, C_{ET}, is defined in such a way that the attenuation by a thickness z of the inhomogeneity is equivalent to the attenuation by a thickness $z \times C_{ET}$ of water. Thus the isodoses are shifted towards the entry surface if $C_{ET} > 1$ and towards greater depths if $C_{ET} < 1$. Nominal values recommended for C_{ET} are: 0.5 for lung, 1.1 for spongy bone, and 1.8 for compact bone.

Figure 1 shows such a shift due to lung tissue, which is proportional to z_L and is approximated closely by the given procedure. The changes of absorbed dose inside the slab of lung tissue and behind this slab need further determination. In accordance with the data in Figure 1 the undisturbed depth-dose curve can be followed through the lung slab, and then the value at the end of the slab can be used for the shift distance $z \cdot C_{ET}$ (indicated by the horizontal broken line). Errors from procedure may be less than about 10%.

The maximum errors in the estimation of absorbed dose inside the slab with lower density, and behind this slab, occur if this slab is made up of air with negligible density. In Figure 2 absorbed dose distributions for air gaps of z_{air} = 1, 3 and 5 cm are given. If the same procedure is used, as described above, about 20% underestimation behind the slab may occur in these extreme cases.

As can be seen from Figure 3, the position of a slab of air inside a water phantom has no influence on the shift of the depth-dose curve. The change in the absorbed dose is more pronounced if the slab is near the entry surface. The approximation, as recom-

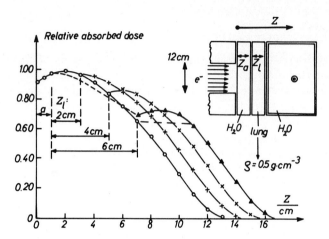

Figure 1. Shift of depth absorbed-dose curves due to slabs of lung tissue of various thickness, z_L, for 25 MeV electrons.

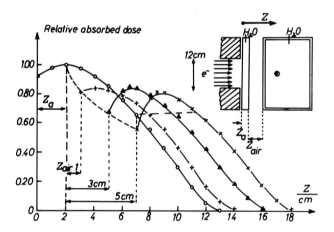

Figure 2. Shift of depth abosorbed-dose curves due to slabs of air of various thickness, z_{air}, for 25 MeV electrons.

mended before, leads to overestimations of less than 20% inside the air gap and underestimations of less than 20% behind the gap. In a real case with lung tissue, these uncertainties are reduced to about 10% and can be tolerated in practical absorbed-dose distributions.

This method of shifting depth-dose curves can also be used as a first approximation for absorbed dose distributions in cases where the inhomogeneities are large uniform slabs of different thickness. The phantom is then cut into strips in the direction of the penetrating electrons and the shift of depth-dose curves is evaluated for the different material inside the separate strips. These depth doses in the different strips are then

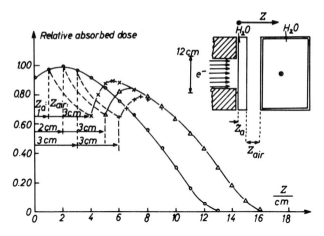

Figure 3. Shift of depth absorbed-dose curves due to a slab of air thickness z_{air} in different depth, z_a, in a water phantom for 25 MeV electrons.

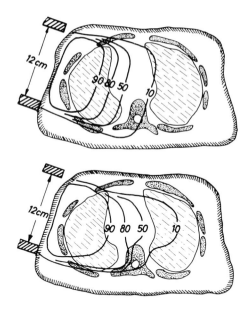

Figure 4. Absorbed dose distribution for 30 MeV electrons measured in water and projected into a body cross-section (above) and corrected absorbed-dose distribution in the same body cross-section using the method of shifting (below).

smoothly connected and can be assumed to give a more realistic picture of the absorbed dose distribution, as demonstrated in Figure 4. Comparative measurements in inhomogeneous phantoms of such a composition have shown that uncertainties of less than about

10% exist. The main effect is due to the large areas of lung tissue. The bone structures, which obviously cannot be thought to be "large uniform slabs of various thickness," only have a relatively small effect on the distribution, as shown in Figure 4. This cannot be assumed for other parts of the human body which are of interest in electron treatment, and therefore in such situations the method of "depth-dose shifting" with suitable C_{ET} values cannot be applied, or can be used only as a very rough approximation (ICRU, 1970).

ABSORBED DOSE DISTRIBUTION BEHIND EDGES OF DIFFERENT MATERIAL

The change of depth absorbed-dose curves behind uniform slabs of a material different from the phantom is due mainly to the difference in density. This change in density affects the linear energy losses of the electrons and thus the range of the electrons.

In contrast to these effects, the change of absorbed-dose distribution behind an edge of a different material is caused mainly by electron scattering, and due to this complicated process, a simple description of this change seems impossible at first glance. But a systematic investigation of dose distributions behind edges of different materials of different shapes has resulted in a suitable method for quantitative description of such absorbed-dose distribution, which will be discussed below for geometrical arrangements with increasing complexity: one edge, stick and cube.

ABSORBED DOSE DISTRIBUTION BEHIND A SINGLE EDGE

If a parallel beam of electrons penetrates solid material, the electrons are scattered in different directions. This effect also influences the shape of the dose-effect curve, but mainly at high primary electron energies (more than about 20 MeV) and at greater depth. For simplicity, in Figure 5 it is assumed that the path of the electrons penetrating the normal phantom material M is a straight line. If a material M' is inserted with a higher mass scattering power (ICRU, 1972), electrons penetrating this layer are scattered, so that behind this slab some electrons are missing. A reduction of the absorbed dose results. These scattered electrons, on the other hand, contribute to the absorbed dose in the normal matter M beyond the edge, and increase the absorbed dose there. At large distances from this edge, in y-direction in Figure 5, there is no influence in M and only a very small influence behind the slab of M', since losses of electrons due to scatter are compensated by equal gains. There is a small reduction in the range of electrons due to the scatter angle away from the penetration direction, which can be neglected in all practical cases. The angle in which these changes of the absorbed dose distribution occur and the magnitude of the changes will be considered below separately.

The most distinct change can be observed directly behind the edge. In Figure 6 an experimental example is given for a lead slab in water. Directly behind the lead there is a distinct reduction, which suddenly jumps to a distinct increase behind the water layer. In larger depths in water this sharp jump in the absorbed dose distribution is spread over a larger region due to electron scattering in water. The side movement of the maxima of reduction and increase, therefore, is dependent only on the mass scattering power of the water and is not dependent on the inhomogenous material at the edge. This is a very important fact, which considerably reduces the complexity of the complete treatment.

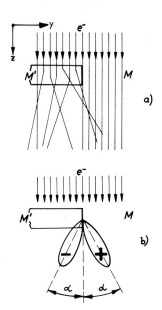

Figure 5. Schematic picture of electron paths behind an edge between different material (a), and their influence on absorbed dose distribution increasing (+) and reducing (-) the original level of absorbed dose (b).

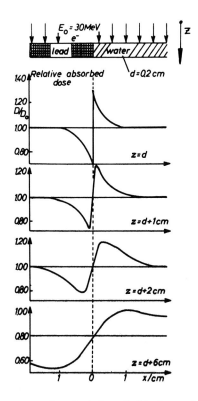

Figure 6. Distribution of relative absorbed dose behind an edge of a lead slab in water measured for fast electrons of $E_0 = 30$ MeV in different depths, z, in water. (x = coordinate perpendicular to the beam direction z)

249

The mean angle, α, for the position of the maxima of reduction and of increase of dose measured which are independent of the composition and thickness of the inhomogeneous material at the edge, as shown in Figure 7. Also a mean angle, β, can be determined outside, where the effect of the edge of the inhomogeneous material is practically negligible.

Such measurements (Fig. 7) have been done with several different materials: lead, silver, carbon, and others positioned at various depths in a water phantom and irradiated with electrons of E_o ranging from 10-30 MeV. It could be proven that both characteristic angles depend only on the mean local electron energy, $\bar{E}$, and are independent of any other physical parameters within the range of experimental uncertainties. The mean local energy, $\bar{E}$, is a function of depth in the phantom, z, and can be calculated by

$$\bar{E} = E_o (1-z/Rp)$$

where E_o is the initial electron energy and Rp the practical range (Harder, 1965; ICRU, 1970). Figure 8 gives these characteristic angles. As can be seen, the angle β is always larger than angle α by a factor of about 2. Therefore only the quantitative relation of α as a function of energy has to be known for a computer program to calculate absorbed dose distribution.

After having determined the geometrical data for the reduction or increase of absorbed dose behind an edge of different material, the magnitude of this change has to be determined. A large number of experiments have been performed with different materials at different depths in a phantom irradiated with electrons of various initial energies, E_o. The maximum change of absorbed dose due to an edge of inhomogeneous material can be expressed by a factor P_{max}, which is defined by

$$P_{max} = \frac{D_{max} - D_o}{D_o}$$

where D_{max} is the absorbed dose at the highest increase or depression, and D_o is the absorbed dose at this point in a homogeneous water phantom. The quantitative values of P_{max} are given in Figure 9 as a function of the local mean electron energy $\bar{E}$ at the position of the inhomogeneity. As can be seen, the influence of an edge increases with increasing electron energy and may reach, in extreme cases of metal (lead) or air changes, up to 30% at electron energies of 30 MeV. For bone in tissue, P_{max} is about half these extreme values.

Very often a rough calculation of such a maximum value would be sufficient for decisions on the suitability of an absorbed dose distribution in an irradiated patient. If a more detailed complete distribution has to be constructed, the decrease of P_{max} with the depth in the tissue has to be taken into consideration. The same is true if the inhomogeneity is more complex than a single edge. Therefore the decrease of P_{max} as a function of m (the distance from the edge) is given in Figure 10 for air in water. Values for bone, again, are about half of these values.

Knowing the two angles α and β (as a function of local mean electron energy as in Fig. 8) and the relative change in absorbed dose (as a function of m and $\bar{E}$ as in Fig. 10), one is able to construct the isodoses behind an edge of inhomogeneous material in the following way. The angles α and β are plotted into a normal absorbed dose distribution

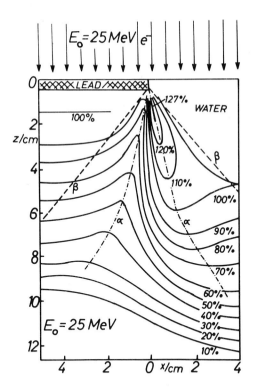

Figure 7. Distribution of absorbed dose behind an edge of a lead slab in water. Definition of the angle α of the maxima of absorbed dose change and of the angle β of negligible change (E_o = 25 MeV).

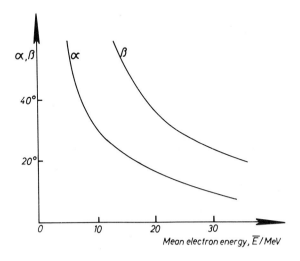

Figure 8. Energy dependence of the angle α (maximum absorbed dose change) and of the angle β (negligible change of absorbed dose) for inhomogeneities in tissue (water).

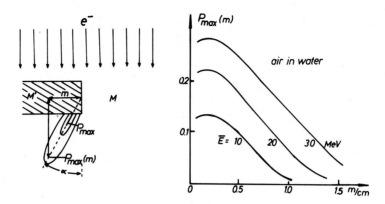

Figure 9. Values for maximum changes in absorbed dose behind edges of inhomogeneous materials in water as a function of local mean electron energy.

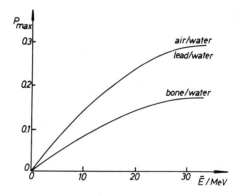

Figure 10. Dependence of the maximum change in absorbed dose with the distance m from the edge of inhomogenous material M' (air in water). Values for p_{max} (m) for bone in water are about half the given data.

in tissue. Behind the inhomogeneous material the shift of absorbed dose is performed as behind large slabs from the side up to the angle β. The values for absorbed dose increase or decrease on the lines of the angle α are inserted and connected with corresponding values on the lines of angle β.

With this procedure quite realistic, absorbed-dose distributions can be predicted with uncertainties of less than 10%.

ABSORBED DOSE DISTRIBUTIONS BEHIND RODS AND CUBES OF INHOMOGENEOUS MATERIAL

This procedure of evaluating the scattering of electrons due to an edge of inhomogeneous material can be applied, even if more than one edge has to be considered. As shown schematically in Figure 11, in the case of a stick or rod of inhomogeneous material, two edges have an effect on the distribution of absorbed dose. In the case of a cube and for approximation of a sphere, four such edges have to be considered. The angles a and β as given in Figure 8 can also be used for these cases. The resulting increase or decrease of absorbed dose behind the inhomogeneity can be calculated with the help of Figure 10. If, for example, a rod-shaped air volume is present with a diameter of 2 times m (where m, as given in Figure 10, is the distance from the edge) then P_{max}, as read from Figure 9, has to be multiplied by a factor of two. For example, the increase of absorbed dose behind a rod-shaped air inclusion of 1 cm diameter for 20 MeV electrons would be $P_{max} = 2 \times 0.16 = 0.32$. Exactly such values can be measured experimentally for such an inhomogeneity.

In the case of cube-shaped inhomogeneities (with sides of, for example, 2 times m) the absorbed dose increase or decrease again can be estimated from Figure 10. But in this case, $P_{max}(m)$ from Figure 10 has to be multiplied by a factor of four. For example, behind an air cube with sides of $2 m = 1.5$ cm, the relative increase of absorbed dose is (see Fig. 10):

$$P_{max} = 4 \times 0.14 = 0.56$$

Again the experimental results are in agreement with this calculated value.

As can be seen from the data in Figure 10, there is a maximal influence on the absorbed dose distribution at a certain diameter of the inhomogeneity of about $2 m = 0.6$ cm. Inhomogeneities of such dimensions sometimes are present in patients and would produce "hot spots" or "cold spots" in the absorbed dose distribution, which may result in changes up to about 50%. They should be considered with care, as their overall effect can be reduced only by multiple-field irradiation. Most of the inhomogeneities present in the human body are larger than one or two centimeters, and the effect on the absorbed dose distribution is reduced, as can be seen from Fig. 10.

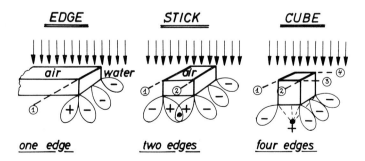

Figure 11. Schematic diagram for the determination of increase and decrease of absorbed dose behind inhomogeneities of various shapes.

In conclusion, one should always consider inhomogeneities in the irradiated body with care and should at least estimate the maximum values for increase and decrease of absorbed dose behind such inhomogeneities. The procedures and quantitative values given here may serve not only for a rough estimation, but also for a more sophisticated calculation of absorbed dose distributions by computers.

REFERENCES

Boone, M.L., P.R. Almond, & A.E. Wright. 1969. High-energy electron dose perturbation in regions of tissue hetergeneity. *Ann NY Acad Sci* 161:214–232.

Dahler, A., A.S. Baker, & J.S. Laughlin. 1969. Comprehensive electron-beam treatment planning. *Ann NY Acad Sci* 161:198–213.

Harder, D. 1965. Energiespektren schneller Elektronen in verschiedenen Tiefen. p. 260. *In* Zuppinger, A. & G. Poretti (eds.) Symposium on High-Energy Electrons – Montreux, 1964. Springer-Verlag, Berlin.

International Commission on Radiation Units and Measurements. 1970. Radiation Dosimetry: Electrons with Initial Energies Between 1 and 50 MeV. Report No. 21. ICRU, Washington, D.C.

Laughlin, J.S., A. Landy, R. Philips, F. Chu, & A. Sattar. 1965. Electron beam treat- *Brit J Radiol* 38:143–147.

Laughlin, J.S., A.R. Landy, R. Philips, F. Chu, & A. Sattar. 1965. Electron beam treatment planning in inhomogeneous tissue. *Radiology* 85:524–531.

Pohlit, W. 1960. Dosisverteilung in inhomogenen Medien bei Bestrahlungen mit schnellen Elektronen. *Fortschr Rontgenstr* 93:631–641.

Pohlit, W. 1969. Calculated and measured dose distribution in inhomogeneous materials and in patients. *Ann NY Acad Sci* 161:189–197.

The Use of High-Energy Electrons in the Treatment of Inoperable Lung and Bronchogenic Carcinoma

W. Schumacher, M.D.,
Professor,
Department of Radiation Therapy
and Nuclear Medicine,
Rudolf Virchow Hospital,
Berlin, Germany

INTRODUCTION

When conventional x-ray therapy with energies up to 250 kV is used for inoperable lung and bronchogenic carcinoma, the survival rates have not been good. The 1-year survival rate without treatment is about 10-15%. Since the advent of megavoltage therapy, the question has arisen of which modality of radiation, high-energy gamma photons, high-energy x-ray photons, or high-energy electrons, is most suitable for irradiation of deeply situated tumors, e.g., in the lungs and mediastinum. The concentration of an adequate dose in the tumor and the sparing of the normal tissues from the effects of radiation are important considerations in the selection of the best modality for irradiation, especially in treatment of lung carcinoma.

Figure 1 shows isodose curves of two opposing fields (8 cm X 10 cm) for 250 kV x-rays in the Alderson phantom; only 50% of the superficial dose is in the center. With a 6000-rad dose in the mediastinal tumor, there is a 9000-rad dose to the spinal cord and a 12,000-rad dose to the skin of the entrance field. To avoid the spinal cord when treating bronchogenic carcinoma with radiation therapy, 80% of all therapists use one direct field ventral and one oblique field dorsal (Heilmann, 1975; Figs. 2-4).

Figures 2, 3 and 4 show isodose curves for cobalt-60, 35-MV photons, and 35-MeV electrons, respectively, for treatment of mediastinal tumors. It can be seen that the isodose distribution changes with the different energies. Only with 30- to 45-MeV elec-

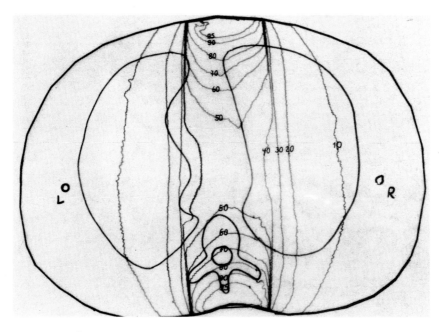

Figure 1. 250 kV x-rays: isodose curves of two opposing fields (8 cm X 10 cm) midthorax in the Alderson phantom. Only 50% of the entrance dose is in the center. With the mediastinal tumor receiving 6000 rads, the spinal cord receives 9000 rads, and the skin, 12,000 rads.

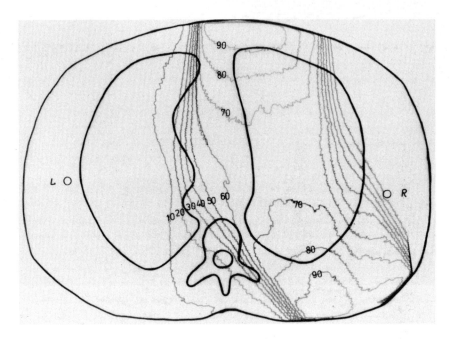

Figure 2. Telecobalt: isodose curves of one direct field ventral and one oblique field dorsal in the Alderson phantom; middle of the thorax; field sizes, 10 cm X 14 cm. Only 60% of the entrance dose is in the center. When the tumor dose is 6000 rads, entrance dose is 10,000 rads, and the dose to the normal lung tissue is 7000-8000 rads.

trons is the homogeneous dose distribution concentrated in the central 5-cm target, with a decreasing dose in the surrounding normal tissues.

Since 1962 we have treated 4,262 cases of lung malignancies with high-energy electrons from a 35-MeV betatron. These patients have been able to tolerate a full course of radiation treatment far better with high-energy electrons than with conventional x-ray therapy. When treated with electrons, the patients gained relief of their symptoms from the tumor growth and gained weight. Their general condition and sense of well-being showed marked improvement during the course of treatment.

We are fortunate in having excellent cooperation and team work with the special lung clinics at Heckeshorn (450 beds) and Havelhöhe (530 beds) Hospitals. All patients receive complete pretreatment examination and evaluation. The examination includes most of the following procedures for each individual case: chest x-rays, tomography, pulmonary function studies, bronchoscopy and pertracheal biopsy of the lymph nodes at the tracheal bifurcation, mediastinoscopy, thoracoscopy, and pleuroscopy. In peripheral tumors transthoracic needle biopsy was mandatory. On the basis of this thorough examination, we feel that we can stage the disease with reasonable accuracy.

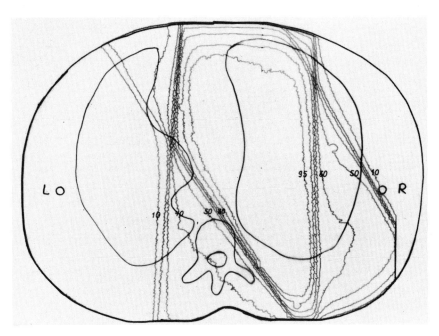

Figure 3. 35-MeV photons: isodose curves of one direct field ventral and one oblique field dorsal in the Alderson phantom; middle of the thorax, field size 10 cm × 14 cm. The 95% isodose curve is in the center. With a dose of 6000 rads in the mediastinum, a wide field, including normal mediastinum and normal lung tissue, receives 6000 rads.

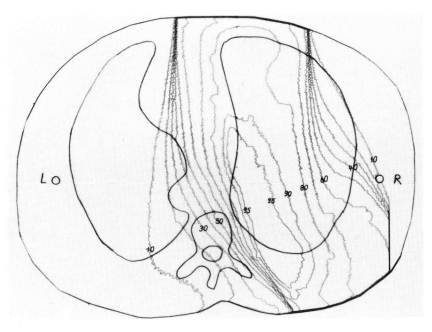

Figure 4. 35-MeV electrons: isodose curves of one direct field ventral and one oblique field dorsal in the Alderson phantom; middle of the thorax, field size 10 cm × 14 cm. With a 6000-rad tumor dose, the normal lung tissue receives only 4,500 rads.

The UICC classification was used for staging, and the 4,262 patients with bronchogenic carcinoma fell into the following stages:

Stage I	396 patients (9.3%)
Stage II	447 patients (10.5%)
Stage III	1,700 patients (39.8%)
Stage IV	1,719 patients (40.4%)
TOTAL	4,262

The median age of this group of patients was 65 years. The following listing gives a more detailed breakdown according to age:

20-39 years	68 patients (1.6%)
40-49 years	170 patients (4.0%)
50-59 years	725 patients (17.0%)
60-69 years	2,221 patients (52.1%)
70-79 years	865 patients (20.3%)
Over 80 years	213 patients (5.0%)
TOTAL	4,262

Of the 4,262 patients treated in our series, only 238 (5.6%) had no pathological proof of cancer; for calculation of the survival statistics, this group was omitted. The pathology of the remaining 4,024 patients with histological or cytological proof of cancer fell into the following categories:

Squamous cell carcinoma	2,122 patients (52.8%)
Oat cell carcinoma	736 patients (18.3%)
Undifferentiated carcinoma	523 patients (13.0%)
Adenocarcinoma	531 patients (13.2%)
Sarcoma	36 patients (0.9%)
Pleuramesothelioma	44 patients (1.1%)
Positive cancer cell, type unknown	32 patients (0.7%)
TOTAL	4,024 patients

DOSE-TIME RELATIONSHIP

At first, for 150 patients we used a conventional irradiation schema, in which three or four individual doses of 300 rads (entrance dose) were given each week to make a weekly tumor dose of 1000 rads per week; this schema was followed for 6–8 weeks to make a total tumor dose of 6000-7000 rads. With this schema, acute radiation pneumonitis, esophagitis, and tracheitis frequently occurred during and after treatment. Late complications included pulmonary fibrosis with increased pulmonary function decompensation, cor pulmonale, and radiation death (16%; Loerbrocks and Schumacher, 1965).

Bestrahlungsrhythmus

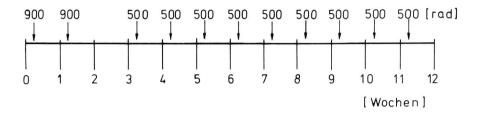

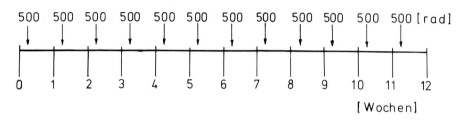

Figure 5. Time-dose relationships used in radiation therapy of patients with bronchogenic cancer and lung tumors. (a) For tumors greater than 6 cm in diameter, vena cava obstruction, abscess, atelectasis. (b) For tumors less than 6 cm in diameter, squamous cell carcinoma.

These serious reactions were an indication that more consideration had to be given to some of the known facts regarding the biological effects of radiation on normal and tumor tissues. Details regarding which radiobiological facts we deemed to be important and the apparent reason for their importance will not be discussed here. As a result of these considerations, we made marked alterations in our time-dose relationship schema (Schumacher, 1965b).

At present our plan is to give 500 rads (entrance dose) per field once a week until a tumor dose of 6000 rads is attained (Fig. 5b). At 3000-rads tumor dose, re-examination and adaptation of the field size to the reduced tumor is made. In approximately 12-14 weeks the patient receives a tumor dose of 6000-7000 rads.

The rate at which the dose is delivered to the tumor and the total tumor dose are adjusted to fit the treatment to the clinical situation. In cases of tumors greater than 6 cm in diameter, tumor abscess formation, vena caval obstruction, or atelectasis, two high individual doses of 900 rads (entrance dose) at an interval of one week (Fig. 5a) are given. This is followed by 2 weeks' rest, and the treatment is then continued at a rate of one treatment per week with 500 rads (entrance dose) until a tumor dose of 5000-7000 rads is reached. Our aim is to give a total tumor dose of 5000 rads for oat-cell carcinoma, 6000 rads for squamous cell carcinoma, and 7000 rads for adenocarcinoma. The good tissue tolerance makes it possible to give 7500 rads for large tumors.

Details regarding the size of the individual prescribed dose, fraction interval, overall radiation time, tumor sensitivity and total tumor dose will be published (Schumacher, in press). Since changing the dose-time fraction from the radiobiological point of view, the 1- and 5-year survival rates showed a marked improvement, as shown in Table 1.

TABLE 1. PATIENTS WITH LUNG AND BRONCHOGENIC CANCER (STAGE III) TREATED WITH CURATIVE DOSES (5000-7000 RADS) OF HIGH-ENERGY (35 MeV) ELECTRONS

	Survival Rates					
	at 1 year			at 5 years		
Year	Number Surviving	Number Treated	(%)	Number Surviving	Number Treated	(%)
1962	84/186		45.2	0/186		0.0
1967	87/152		57.2	9/152		5.9
1969	101/165		61.2	14/165		8.1

INDICATIONS FOR THE USE OF HIGH-ENERGY ELECTRONS

It has been previously stated by some authors (Becker and Shubert, 1961; Brobowitz et al., 1961) that radiation therapy is contraindicated for cases of bronchogenic carcinoma with complications of tuberculosis, abscess formation, or severe vena caval obstruction in association with bronchial stenosis and abscesses. From our experience in using high-energy electrons with curative dosages, we can say that these complications should no longer be considered contraindications to radiation treatment. On the contrary, only when radiation therapy, combined with appropriate antibiotic therapy, is used, is there an opportunity to reduce the tumor obstruction of the bronchus and thus allow proper drainage of the abscess and subsequent regression.

Tuberculosis. The presence of active or inactive tuberculosis in association with lung cancer is also no longer a contraindication to curative radiation therapy (Broll et al., 1972). However, it is of the utmost importance that the patient receive adequate and continuous antituberculosis chemotherapy during the radiation therapy and for some months afterwards. We have treated 94 patients (92 men and 2 women) with bronchogenic carcinoma and tuberculosis. Of these, the diagnosis of tuberculosis for 19 of the patients was confirmed by culture and radiographic evidence of cavity formation. In 86.3% of the patients the tuberculous process was bilateral with, or located on the same side as, the carcinoma. In all patients there was regression of the tumor from radiation therapy and regression of the tuberculosis from the antituberculosis chemotherapy. The regression of the cavity was not influenced by the curative radiation therapy. No patient with active or inactive tuberculosis, who received antituberculosis chemotherapy, showed aggravation of the tuberculosis.

TABLE 2. FREQUENCY OF COMPLICATIONS IN SERIES OF 4,024 PATIENTS WITH HISTOLOGIC AND CYTOLOGIC PROOF OF LUNG AND BRONCHOGENIC CARCINOMA

Complication	Number Patients	Percentage of Total (%)	
Tuberculosis			
Active tuberculosis with positive tubercle bacilli	68	2.7	
Active and inactive tuberculosis	188	7.5	
			10.2
Hemoptysis	384	15.3	
Tumor abscess formation	56	2.2	
Vena caval obstruction	68	2.7	
			20.2
TOTAL		30.4	

In two patients with inactive tuberculosis, who did not receive prophylactic chemotherapy, the tuberculosis was exacerbated; this indicates that all patients with tuberculosis, active or inactive, need prophylactic antituberculosis chemotherapy during radiation treatment. The one-year survival rate for patients with bronchogenic carcinoma complicated with tuberculosis was 41% (Broll et al., 1972).

Case Report. A 71-year-old patient with advanced, active tuberculosis and cavity formation with tubercle bacilli was treated with antituberculosis chemotherapy (Fig. 6). During treatment he developed a squamous cell carcinoma of the left bronchus which was confirmed by bronchoscopy and histologic examination (Fig. 7). This patient was irradiated with high-energy (35 MeV) electrons with a direct, 10 cm X 14 cm field ventral and a left, oblique, 10 cm X 14 cm field dorsal and received a 6000-rad dose to the tumor ($T_2N_1M_0$). Five months later there was full regression of the tumor and further regression, with no reaggravation, of the tuberculosis (Fig. 8).

Tumor Abscess or Poststenotic Abscess Formation. We have treated 53 patients with inoperable bronchogenic carcinoma complicated by tumor necrosis and formation of abscesses up to 8-10 cm in diameter. In 38 patients there was marked regression of the abscess during treatment. In 15 cases there was no further progression of the abscess, and the fever subsided. These favorable results with radiation therapy are in sharp contrast to those obtained in a group of 11 patients who received *only cytotoxic agents* (Endoxan) and antibiotics. In that group *no regression* of the abscess was observed, and in two patients the tumor abscess increased in size. Figure 9 shows a radiogram of a 54-year-old patient with histologically confirmed, squamous-cell carcinoma with a tumor abscess in the upper lobe of the right lung. This patient was irradiated with high-energy (35 MeV) electrons via two, opposing, 8 cm X 10 cm fields, once a week with a 500-rad entrance dose, for a total tumor dose of 6000 rads in 14 weeks. Figure 10, a radiogram of the same patient 5 months after treatment, shows regression of the tumor abscess.

Hemoptysis. Hemoptysis from bronchogenic carcinoma is an indication for the use of radiation therapy. We have treated 62 patients with marked hemoptysis; in 56 of these

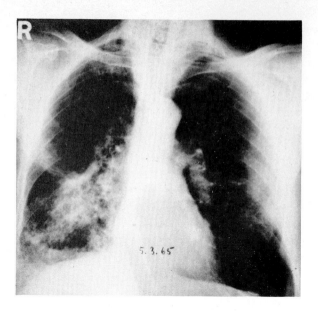

Figure 6. Chest radiogram of a 71-year-old patient with extensive active tuberculosis with cavity formation and positive sputum culture.

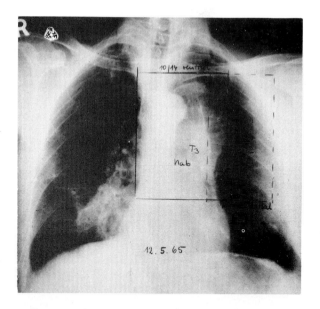

Figure 7. Chest radiogram of the same patient 2 months later, with atelectasis of the upper left lobe.

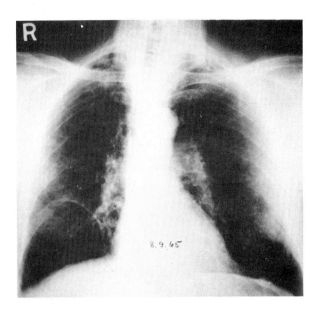

Figure 8. Chest radiogram of the same patient 2 months after radiation therapy with high-energy electron and antituberculosis chemotherapy. The tumor has fully regressed, and the tuberculosis has regressed further, without aggravation during the radiation treatment.

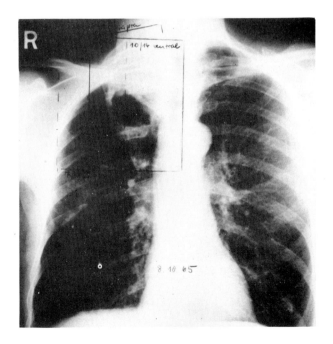

Figure 9. Radiogram of a patient with histologically proven squamous-cell carcinoma with tumor abscess and sputum culture positive for tuberculosis.

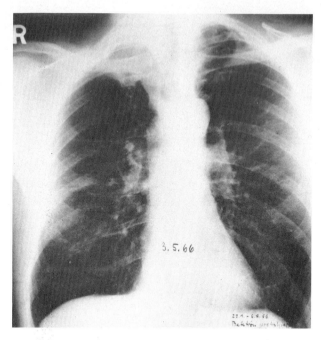

Figure 10. The same patient 4 months after radiation treatment with high-energy electrons. The tumor abscess and tuberculosis have regressed.

patients the symptoms subsided during radiation treatment. The incidence of hemorrhagic death was not increased by the radiation therapy.

Severe vena caval obstruction. The distressing symptoms of the superior vena caval syndrome can be relieved most easily by treatment with high-energy electrons. For all cases of vena caval obstruction, we now give one 900-rad (entrance dose) treatment per field in one week (see Fig. 5a), and after a 2-week interval the tumor is irradiated with doses of 500 rads once per week, for a total tumor dose of 6000 rads in 12-14 weeks. With this schema, the patient is relieved of the severe symptoms of the vena caval obstruction one week after receiving the first doses of radiation.

Case Report. An example of this type of problem and its treatment is a 60-year-old patient with histologically proven, small-cell carcinoma with a large (10 cm X 12 cm) tumor, who had a severe vena caval obstruction and required oxygen administration for marked dyspnea (Figs. 11-13). We treated this patient with two, 1,200 rad, entrance doses of 35-MeV electrons through a direct, 14 cm X 14 cm field ventral and an oblique, 14 cm X 14 cm field dorsal, with a one-week interval between irradiations. We followed these high single doses with reduced weekly doses for a total tumor dose of 4,500 rads in 12 weeks. Twelve years later the patient is still alive, with no recurrence of the tumor, no symptoms of fibrosis, and no metastases. Figures 14 and 15 are tomograms of the patient before and after radiation treatment.

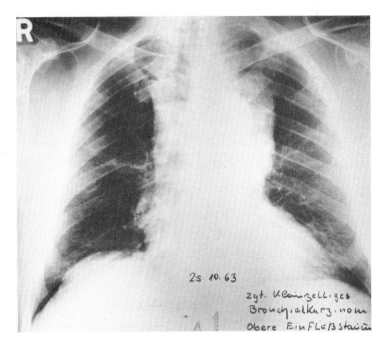

Figure 11. Chest radiogram of a 60-year-old patient with histologically proven, small-cell carcinoma (tumor size: 10 cm X 12 cm), with a severe vena caval obstruction.

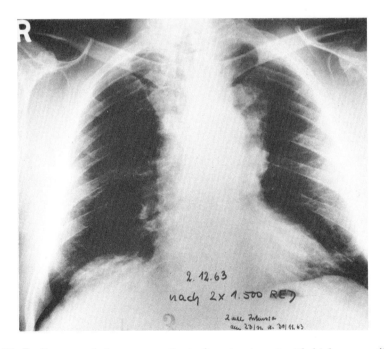

Figure 12. Radiogram of the same patient after treatment with high-energy (35 MeV) electrons in two doses of 1,200-rads (at the maximum of the depth-dose curve with a one-week interval.

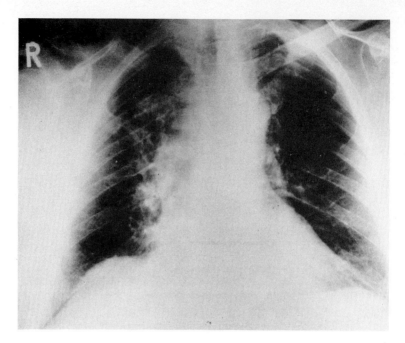

Figure 13. Radiogram of the same patient 12 years after treatment (total tumor dose: 4,500 rads). There has been no recurrence of the tumor or metastasis.

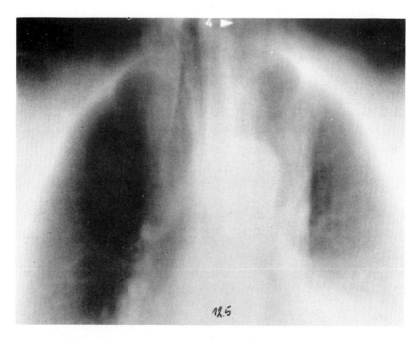

Figure 14. Tomogram of the patient before radiation therapy.

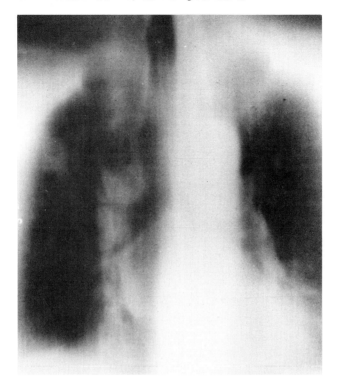

Figure 15. Tomogram of the patient 12 years after treatment.

We saw tumor regression in all the patients we treated curatively with high single doses of radiation followed by reduced weekly doses. In the treatment of small- or oat-cell carcinoma, we treat the large primary tumor with 900 rads once a week for 2 weeks followed by 500-rad weekly doses after a 2-week interval (see Fig. 5a). The supraclavicular right and left lymph nodes are also irradiated, tangentially from 8 cm X 10 cm ventral and dorsal fields, once a week with 500-rad (entrance) doses, to a total dose of 3000 rads. (The lymph nodes are irradiated on a different day of the week from the day on which the primary tumor is irradiated.)

Figures 16 and 17 show an example of a patient treated in such a manner. These are chest radiograms of a 47-year-old patient with a 12 cm X 14 cm tumor (T_4N_2) from histologically proven, small-cell carcinoma (Figs. 16a and b). This patient received a total tumor dose of 6000 rads in 14 weeks with 35-MeV electrons. After treatment the tumor had fully regressed (Figs. 17a and b). The patient died from metastasis in the pancreas, but with no tumor in the irradiated region.

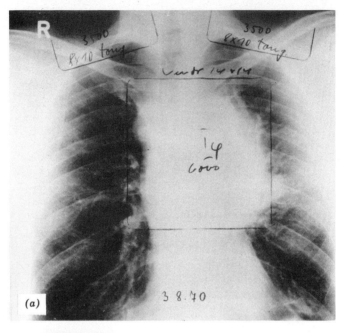

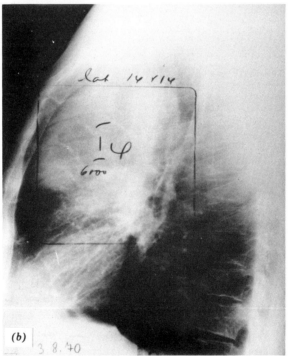

Figure 16 a and b. Radiogram of a 47-year-old patient with small-cell carcinoma.

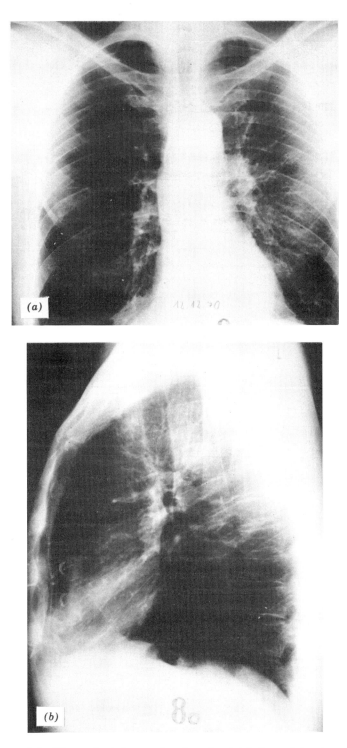

Figure 17 a and b. The patient 4 months after the initiation of treatment with high-energy electrons.

PALLIATIVE RADIATION THERAPY

In administering palliative or symptomatic therapy of rapidly progressing cases of bronchogenic carcinoma, we start with high individual doses of 900 rads (Fig. 5a), to a total dose of 3,000 rads. Given in this manner, the doses are well tolerated by the lung tissue. In some cases it was possible to give curative doses (Figs. 11-15).

Of the 4,024 patients treated, 1,122 (27.8%) were irradiated with palliative tumor doses. These patients were primarily in an advanced stage and had developed metastases and complications from their other diseases.

Complications. Since we adopted the time-dose relationship shown in Figure 5, only 50% of our patients have shown radiographic signs of radiation reaction. In 49% (1,422) of the patients, there were no signs of such reactions, 32% (928) showed mild reactions, and 19% (552) showed reactions with fibrosis and shrinking of the irradiated part of the lung. Pneumonitis and pulmonary fibrosis are not severe complications of radiation therapy with high energy-electrons. In the total care of the patient, cooperation between the radiotherapist, lung clinician, and the general practitioner is of prime importance. During, and for at least two months after, radiation therapy the patient must be carefully evaluated at weekly intervals. Elevation of the body temperature and the sedimentation rate (over 100 mm Westergreen) should make one suspect a diagnosis of radiation pneumonitis. A chest radiogram should be taken immediately when this diagnosis is suspected. A few days of therapy with the appropriate antibiotics, cortisone and tanderil will usually control the condition.

An example of such a problem is shown in Figures 18-21. The 68-year old patient had histologically proven, undifferentiated-cell carcinoma (T_3N_1); a pretreatment radiogram is shown in Figure 18. The patient received a total tumor dose of 6,000 rads with 35-MeV electrons over a period of 14 weeks. Figure 19 shows a radiogram taken one week after the completion of this treatment. Two weeks after the completion of the radiation treatment, the patient developed acute radiation pneumonia (Fig. 20). The patient was immediately treated for the pneumonia with antibiotics, cortisone and tanderil, and 10 days later, the pneumonia had fully regressed (Fig. 21).

Application of the proper treatment, as indicated above, resulted in clinical improvement within a few days and control of the complication. If good treatment results are to be expected, it is mandatory that the patient be carefully observed in the months following completion of the treatment. Follow-up bronchoscopies and biopsies after radiation treatment were made in 30% of our patients. In 70.3% of these, the pathological examination of the biopsy specimen showed no residual cancer.

It is of interest to compare the survival time of our patients with inoperable bronchogenic carcinoma treated with high-energy electrons, with others results reported in the literature (Franke et al.; see Table 3). The one-year survival rate with cobalt-60 radiation therapy in a small number of patients gives a survival rate of about 27-34%.

With high-energy (35 MeV) electrons giving a total tumor dose of 5000-6000 rads, in 1,736 patients, Stages I-III, the one-year survival of 1,093 patients (no selected cases) was 62.9%.

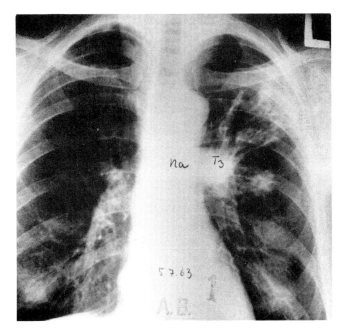

Figure 18. Chest radiogram of a 68-year-old patient with histologically proven, undifferentiated, cell carcinoma of the lung.

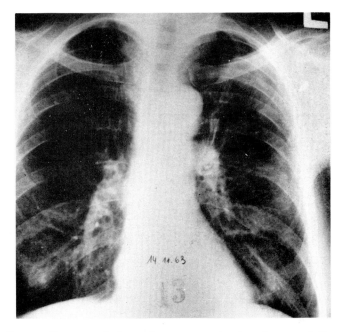

Figure 19. Chest radiogram of the patient one week after the completion of treatment with 35-MeV electrons (6,000 rads tumor dose).

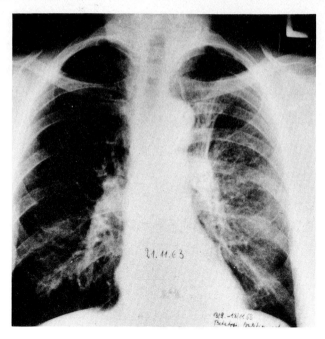

Figure 20. Chest radiogram of the patient 2 weeks after completion of radiation treatment; acute radiation pneumonia has developed.

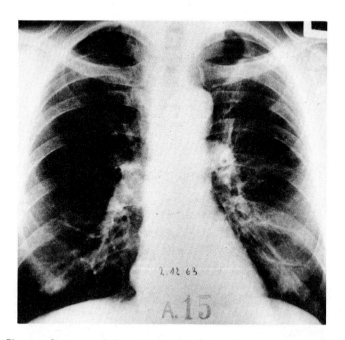

Figure 21. Chest radiogram of the patient 10 days after treatment of the pneumonia with antibiotics, cortisone and tanderil; the inflammation has fully regressed.

TABLE 3. SURVIVAL RATE AFTER RADIATION TREATMENT OF INOPERABLE LUNG AND BRONCHOGENIC CARCINOMA

Reference	Year	Type of Radiation	Number of Patients	Survival Rate (%)				
				1 yr	2 yr	3 yr	4 yr	5 yr
Kutz	1958	Co-60	173	24	7			
Lofstrom	1959	Co-60	100	19				
Guerin	1960	Co-60	120	34	12			
Valdagni	1960	Co-60	287	32.4	10.2			
Kuttig et al.	1962	Co-60	406	25	10	6		
Birkner, Hinz	1963	Co-60	190	34				
Burr et al.	1963	Co-60	862	26.9				
Schnepper	1963	Co-60	112	27.6				
Trial et al.	1963	Co-60	111	30				
Cochi	1964	31 MeV x-ray	255	18		4		2
Pedoja, Rigat	1964	Co-60	74	18.9				
Sarasin, Sayegh	1964	Co-60	74	ca. 8				
Smith et al.	1964	Co-60	862	26	11	8		
Beling and Einhorn	1965	Co-60	138	22	5			
Brady et al.	1965	2 MeV x-ray	28	53	29	12		
Guttmann (post thoracotomy inop.)	1965	2 MeV x-ray	95	58	28	18	11	7
Guttmann (Clin. inop.)	1965	2 MeV x-ray	150	40	13	7	5	2.5
Jacob et al.	1965	Co-60	36	14				
Mustakallio	1965	Co-60	51	25	16			
Pierquin et al.	1965	22 MeV x-ray	355	30		10		
Hess and Buchelt	1966	Co-60	151	32	13	7	7	3

TABLE 3. SURVIVAL RATE AFTER RADIATION TREATMENT OF INOPERABLE LUNG AND BRONCHOGENIC CARCINOMA (Cont'd.)

Reference	Year	Type of Radiation	Number of Patients	Survival Rate (%)				
				1 yr	2 yr	3 yr	4 yr	5 yr
de Jong and Renner	1966	Co-60	103	25	10			
Deeley and Singh	1967	8 MeV x-ray	513	36	15	8	7	6
Franke et al.	1967	Co-60	144	24	6	2		
Holsti	1967	Co-60	102	26	7			
Schnepper and Vielberg	1967	Co-60	121	34	12	8	5	5
Cook et al.	1968	2 MeV x-ray	64	20	3			
Eichhorn and Lessel	1968	Co-60	241[a]			6	6	6
			162[b]			1		
Ott	1968	Co-60	390	7	2	1		
Fernholtz and Müller	1969	Co-60	124	30	7	2	0.6	
Taskinen	1969	Co-60	51	29				
Teschendorf and Bleher	1969	42 MeV electrons	136	32				
Schumacher	1962-1969	35 MeV electrons	1736	62.9	28.9	15.8	9.0	8.0

[a]Curative
[b]Palliative

In comparing the statistical results in the literature, it is important to know some other data:

1. The staging of cancer and its composition, since in Stages I or II, the 1-year survival rate—depending on the tumor growth rate—is longer than the survival rate in patients with stage III cancer. In our 2,223 patients with inoperable bronchogenic carcinoma, 80% were cases of Stages III and IV (Table 4).

2. Some of the statistical results in the literature include some cases of not histologically confirmed carcinoma. These cases give the possibility of a better 5-year survival rate, especially when the number of 5-year survivals is very small. The statistics for our patients include only cases of histologically and cytologically (Papanicolaou Grade V) proven cancer.

3. The survival rate also depends on the median age of the patients. In survival statistics in the literature, the median age was in the range of 50-60 years. In our cases, the median age was about 65 years. The death rate from diseases other than lung cancer is not calculated into the above survival rate.

4. From the small number of patients reported it can be concluded that they have been mostly preselected. A selection can be assumed from the proposed contraindication in cases of simultaneous presence of tuberculosis and other complications. In our statistics we have included unselected patients with a high risk from complications, such as active and inactive tuberculosis (10%), hemoptysis (15.3%), tumor abscess formation (2.2%) and severe vena caval obstruction (2.7%).

Radiation doses in the range of 5,000-7,000 rads in our fractionation schema with high-energy electrons also give better results in Stage IV survival rates.

The discrepancy of staging after surgical intervention or before radiotherapy is one reason for the better results in the statistics of the operable cases. When comparing with the irradiated cases, we have tried to calculate the error of staging on the basis of radiographic examination alone, in comparison with that done at the time of surgical exposure using post-mortem studies. The results are evident from a comparison of the Tables 4 and 5. The 5-year survival rate in these patients is shown in Table 6. Including all patients, Stages I-III, the one-year survival rate for 1,736 patients was 62.9%.

The efficiency of curative radiation treatment can only be based upon radiation treatment in Stages I-III (Table 7). In 55.8% the tumor was central, and in 44.2%, in the periphery. The 5-year survival rate according to the histological type of tumor cells (Table 8) shows that in adenocarcinoma, radiation therapy with curative doses is indicated. Properly planned radiation therapy with high-energy electrons gives a marked increase in survival time, but, in addition, the comfort and well-being of the patient is often so much improved that many patients are able to continue their normal daily life pattern for a much longer period of time. The comparatively favorable survival rate of patients with adenocarcinoma indicates that these patients should also be treated with curative doses of high-energy electrons.

SUMMARY

A better radiation dose distribution can be obtained in the tumor and the regional lymph nodes in the mediastinum with the use of high-energy electrons than with high-energy x-rays or cobalt-60 therapy. In our department, treatment with high-energy

TABLE 4. INOPERABLE LUNG AND BRONCHOGENIC CARCINOMA: ONE-YEAR SURVIVAL RATE[a]

Year	Stage I		Stage II		Stage III		Stage IV		Total	
	No.[b]	%	No.	%	No.	%	No.	%	No.	%
1962-64	17/24	70.8	22/43	51.1	84/186	45.2	24/181	13.3	147/434	33.3
1965	12/21	57.1	21/44	47.7	51/101	50.4	15/119	12.6	99/285	34.7
1966	16/21	76.2	17/36	47.2	57/99	57.6	18/146	12.3	108/302	35.7
1967	25/30	83.3	31/55	56.3	87/152	57.2	26/61	42.6	169/298	56.7
1968	32/45	71.1	27/41	65.8	88/152	57.8	15/37	40.5	162/275	58.9
1969	48/61	78.7	22/32	68.7	101/165	61.2	26/57	45.6	197/315	62.5
1970	25/29	86.2	8/10	80.0	62/78	79.4	47/167	28.1	142/284	50.0
1971	20/23	86.9	12/14	85.7	51/65	78.4	20/102	19.6	103/204	50.4
1972	7/8	87.5	12/17	70.5	56/67	83.5	33/165	20.0	108/257	42.0
1973	7/8	87.5	14/15	93.3	61/94	64.8	29/131	22.1	111/248	44.7
TOTAL	209/270 (77.4%)		186/307 (60.5%)		698/1159 (60.2%)		253/1166 (21.9%)		1346/2902 (46.3%)	

[a] A series of 2,902 patients treated from 1962 to 1973 with curative tumor doses (5000-7000 rads) in 12-14 weeks with high-energy (35-45 MeV) electrons for histologically or cytologically confirmed bronchogenic and lung carcinoma.

[b] Number Surviving/Number Treated

TABLE 5. INOPERABLE LUNG AND BRONCHO-GENIC CARCINOMA: ONE-YEAR SURVIVAL RATE[a]

Year	Survival Rate No.[b]	%
1962-64	123/253	48.6
1965	84/166	50.6
1966	90/156	57.6
1967	143/237	60.3
1968	147/238	61.7
1969	171/258	66.2
1970	95/117	81.1
1971	83/102	81.3
1972	75/92	81.5
1973	82/117	70.0
TOTAL	1,093/1,736	62.9

[a] A series of 1,736 patients treated from 1962 to 1973 with curative tumor doses (5000-7000 rads) of high-energy (35-45 MeV) electrons for histologically or cytologically confirmed bronchogenic and lung carcinoma, Stages I-III.

[b] Number Surviving/Number Treated

TABLE 6. FIVE-YEAR SURVIVAL RATE (FROM 1969) FOR 166 PATIENTS, WITHOUT STAGE DISCREPANCY FROM POST-MORTEM STUDIES

Stage	5-Year Survivals Number[a]	%
Stage I	7/21	33.3
Stage II	6/21	28.5
Stage III	9/124	7.2

[a] Number Surviving/Number Treated

TABLE 7. SURVIVAL RATES, STAGES I-III

Years	Survival Rate	
	Number[a]	%
1	1,736/1,093	62.9
2	1,619/468	28.9
3	1,527/242	15.8
4	1,425/128	8.9
5	1,308/106	8.1

[a]Number Surviving/ Number Treated

TABLE 8. FIVE-YEAR SURVIVAL RATE ACCORDING TO HISTOLOGIC TYPE OF TUMOR CELL

Type	5-Year Survivals	
	Number[a]	%
Squamous cell carcinoma	51/738	6.9
Oat cell carcinoma	18/229	7.8
Undifferentiated carcinoma	10/152	6.5
Adenocarcinoma	23/154	14.9
Sarcoma	2/11	18.1
Pleuramesothelioma	1/14	7.1
Positive cancer cells, type unknown	1/10	10.0

[a]Number Surviving/Number Treated

electrons has resulted in a one-year survival rate of 62.9% and a 5-year survival rate of 8% for patients with inoperable lung cancer. Proper treatment with high-energy electrons is well tolerated by patients with malignant disease of the lung. A close cooperation between the radiation therapist, lung clinician and general practitioner throughout the course of radiation therapy and during the months following treatment is essential if good results are to be expected. The complication of radiation pneumonitis can be controlled; lung abscess, moderately severe hemoptysis or tuberculosis should *not* be considered as contraindications to radiation therapy. In all cases we saw tumor regression. No spinal or cardiac complications were observed. The reason for our better survival rates is the good dose distribution and concentration of high-energy electrons. Our dose fractionation, from the biological standpoint, gives a good tolerance without severe side reactions and late complications.

REFERENCES

Abe, M., M. Takahashi, Y. Onoyama, H. Sai, T. Nishidai, & S. Oshima. 1971. Radiotherapy of carcinoma of the lung. *Nippon Acta Radiol* 31:825–832 (Japanese).

Bassi, P. & G. Sanna. 1970. Considerazioni su 50 casi di tumori polmonari centrali allo stadio T_2 trattati con telecobaltoterapia. *Radiobiol Radioter Fis Med* 25:131–140 (Italian).

Becker, J. & G. Schubert. 1961. Die Supervolttherapie; Grundlagen, Methoden und Ergebnisse der Therapie mit energiereichen Teilchen und ultraharten Strahlen. Thieme Verlag, Stuttgart.

Beling, U. & J. Einhorn. 1965. Radiotherapy for carcinoma of the lung. *Acta Radiol [Ther] (Strockholm)* 3:281–286.

Birkner, R. & G. Hinz. 1963. [Supervolt therapy of bronchial carcinoma.] *Radiologe (Berlin)* 3:182–187 (German).

Bobrowitz, I.D., M. Elkin, J.C. Evans & A. Lin. 1961. Effect of direct irradiation on the course of pulmonary tuberculosis (using cancerocidal doses). *Dis Chest* 40:397–406.

Brady, L.W., P.A. Germon, & L. Cander. 1965. The effects of radiation therapy on pulmonary function in carcinoma of the lung. *Radiology* 85:130–134.

Broll, L., K.L. Radenbach & W. Schumacher. 1972. Die Indikation zur Strahlentherapie beim Bronchialcarcinom in Kombination mit Lungentuberkulose. Probleme und Behandlungsergebnisse. Vortrag Ges. f. Tuberkulose u. Lungenkrankheiten (German).

Bublitz, G. & R. Labitzke. 1968. Zur Behandlung des Bronchialkarzinoms unter besonderer Berücksichtigung der Röntgen- und Telekobalttherapie. *Strahlentherapie* 135:513–523 (German).

Burr, R.C., E.N. MacKay & A.H. Sellers. 1963. Radiation therapy in the treatment of carcinoma of the lung. *Canad Med Ass J* 88:1181–1184.

Cocci, U. 1964. Was erreicht die Strahlentherapie beim Bronchuscarcinoma? Klinische-statistische Untersuchung von 591 Patienten. *Radiol Clin (Basel)* 33:93–110 (German).

Cook, J.C., H.J. West & J.W. Kraft. 1968. The treatment of lung cancer by split-dose irradiation. *Amer J Roentgen* 103:772–777.

Deeley, T.J. & S.P. Singh. 1967. Treatment of inoperable carcinoma of the bronchus by megavoltage X rays. *Thorax* 22:562–566.

Eichhorn, H.J. & A. Lessel. 1968. Spätresultate nach Telekobalttherapie bei histologisch gesichertem inoperablem Bronchialkarzinom. *Strahlentherapie* 136:411–413 (German).

Felci, U. 1969. La nostra esperienza della telecobaltoterapia del carcinoma inoperabile del polmone. *Radiol Med (Torino)* 55:568–574 (Italian).

Felci, U., F. Milani, R. Musumeci, G. Viganotti & R. Zucali. 1971. Telecobaltoterapia delle neoplasie polomonarie inoperabili: 276 casi. *Radiol Med (Torino)* 57:56–62 (Italian).

Fernholtz, H.J. & G. Müller. 1969. Ergebnisse und Komplikationen der Telecobalttherapie beim Bronchialkarzinom. *Strahlentherapie* 137:381–392 (German).

Franke, H.D., H.P. Haug & G. Stephan. 1967. Anwendung des TNM-Systems bei inoperablen Bronchuskarzinomen zum Vergleich der Ergebnisse nach Röntgen- und Telekobalttherapie. *Strahlentherapie* 132:161–172; 334–351 (German).

Franke, H.D. & H.W. Kunstmann. 1970. Strahlentherapie des Bronchialcarcinoms. *Internist (Berlin)* 11:334–343 (German).

Guerin, R.A. & M.T. Guerin. 1960. [Radioisotopes.] *Vie Med* 41:161–164 (French).

Guthrie, R.T., J.J. Ptacek & A.C. Hass. 1973. Comparative analysis of two regimens of split-course radiation in carcinoma of the lung. *Amer J Roentgen* 117:605–608.

Guttmann, R.J. 1965. Results of radiation therapy in patients with inoperable carcinoma of the lung whose status was established at exploratory thoracotomy. *Amer J Roentgen* 93:99–103.

Heilmann, H.P. Ergebnisse der Strahlentherapie des Bronchuskarzinoms. Vort. Röntgengesellschaft, Tagung 1.–3. Mai 1975, Berlin, *In Press*.

Holsti, L.R. 1967. Roentgen and tele-cobalt therapy of cancer of the lung. *Acta Radiol [Ther] (Stockholm)* 6:65–73.

Jacob, P., J.S. Abbatucci, J. Robillard & A. Mouchel. 1965. Résultats comparés du traitement des cancers pulmonaires inopérables par télécobalthérapie et radiothérapie à 200 kV. *J Radiol Electrol* 46:465–469 (French).

Kuttig, H., J. Becker & H.J. Frischbier. 1962. [Experiences and results of radiotherapy in bronchial carcinoma.] *Strahlentherapie* 118:326–340 (German).

Kutz, E.R. 1958. Intensive cobalt-60 teletherapy of lung cancer. *Radiology* 71:327–335.

Loerbroks, B. & W. Schumacher. 1965. Behandlungsergebnisse und intrathorakale Strahlenreaktionen bei der Therapie des Bronchuscarcinoms mit schnellen Elektronen. I. Behandlungsergebnisse. II. Strahlenreaktionen. *Beitr Klin Tuberk* 131:131–143, 144–167 (German).

Mustakallio, S. 1965. Vergleichende Untersuchungen mit der Tele-Kobalt-60-Therapie und der konventionellen Röntgenbehandlung bei Lungen- und Ösophaguskarzinom. *Röntgenblätter* 18:537–539 (German).

Ott, A. 1968. Ergebniss der Telekobalttherapie beim inoperablen Bronchuskarzinom. *Strahlentherapie* 136:6–10 (German).

Palojoki, A. & V. Nikkanen. 1975. Bronchoscopical changes after megavoltage in patients with bronchial carcinoma. *Excerpta Med (Radiol)* 32:2240 (Abstract).

Pedoja, G. & L. Rigat. 1964. Risultati del trattamento dei tumori maligni del polmone mediante telecobaltoterapia. *Radiol Med (Torino)* 50:639–652 (Italian).

Pierquin, B., P. Gravis & X. Gelle. 1965. Etude de 688 cas de cancers bronchiques traités par téléradiothérapie (200 kV et 22 MV). *J Radiol Electr* 46:201–216 (French).

Rado, M. 1974. Zur Strahlentherapie des Bronchialkarzinoms. *Röntgenblätter* 27:194–198 (German).

Sarasin, R. & C. Sayegh. 1964. Traitement du cancer pulmonaire par le télécobalt. *Radiol Clin (Basel)* 33:143–166 (French).

Schnepper, E. & H. Vielberg. 1967. Ergebnisse der Kobalt-60-Teletherapie des Bronchialkarzinoms. *Strahlentherapie* 133:176–183 (German).

Schumacher, W. 1965. Zweijährige klinische Beobachtungen bei der Anwendung schneller Elektronen des 35-MeV-Betatron. Dtsch. Röntgenkongress 1964, Teil B. *Sonderbände zur Strahlentherapie* 61:74–81 (German).

Schumacher, W. 1965. Die Änderung des Bestrahlungsrhythmus. p. 258–261. *In* Symposium on High-Energy Electrons, Montreux, Switzerland, Sept. 1964. (Zuppinger, A. & G. Poretti, eds.) Springer Verlag, New York (German).

Schumacher, W. 1967. Neue strahlenbiologische Erkenntnisse zur Verbesserung der Strahlentherapie. Dtsch. Röntgenkongress 1966, Teil B. *Sonderbände zur Strahlentherapie* 64:122–129 (German).

Schumacher, W. 1972. Nutzbarmachung neuer Erkenntnisse über die Fraktionierung bei der Bestrahlung bösartiger Tumoren für die Praxis. *Röntgen-Berichte* 1:91–100 (German).

Taskinen, P.J. 1969. Large dose fractions in telecobalt radiotherapy. A report on the treatment of inoperable lung cancer. *Strahlentherapie* 137:522–529.

Teschendorf, W. & E.A. Bleher. 1969. Erste Erfahrungen mit einem 42 MeV Betatron bei der Therapie des Bronchialcarcinoms. *Röntgenblätter* 22:432–442 (German).

Trial, R., R. Roze, D. Uzzan & J. Chebat. 1963. [Cobalt teletherapy of malignant tumors. (Technic, complications and results.)] *Sem Hop Paris* 39:730–739 (French).

Valdagni, C. 1960. [Five years of telecobalt therapy.] *Minerva Fisioter* 5:45–68 (Italian).

Wideröe, R. 1969. Tumor therapy with high-energy ionizing radiation. *Kerntechnik* 11:312–317.

Wideröe, R. 1968. Neutronenstrahlung in der Medizin. *Fortschr Röntgenstr* 108:659–668 (German).

Wideröe, R. 1971. Various examples from cellular kinetics showing how radiation quality can be analysed and calculated by the two-component theory of radiation. p. 311–330. *In* Symposium on Biophysical Aspects of Radiation Quality, 3rd, Lucas Heights, Australia, 1971. Vienna. (IAEA-SM 145/44).

Wideröe, R. 1971. Quantitative and qualitative aspects of radiobiology and their significance in radiation therapy. *Acta Radiol [Ther] (Stockholm)* 10:605–624.

Wideröe, R. 1970. A discussion of the oxygen effect based upon the two-component theory of radiation. *Radiol Clin (Basel)* 39:20–31.

Wideröe, R. 1969. A critical review of radiation therapy using hyperbaric oxygen treatment. *Fortschr Röntgenstr* 110:738–744.

Wideröe, R. 1968. Analytische Untersuchungen von Bestrahlungsprogrammen mit ver schiedenen Strahlenqualitäten. *Stud Biophys* 8:659 (German).

Wideröe, R. 1973. Fractionation schemes. Calculation of a radiobiological model and of clinical investigations on tumor destruction. Lecture, Radiology, 13th International Congress, Madrid.

Wideröe, R. The Ellis formula. A study in applied radiobiology. International Congress of Radiation Research, 5th, Seattle, Washington, July 1974, *In Press*.

Telecentric Rotation with Electron Beams

Dieter Fehrentz, Ph.D.,
Universitäts-Strahlenklinik,
Heidelberg, Germany

When depth-dose curves for electrons of a 42-MeV betatron (Schittenhelm et al., 1965) with field homogenized by scattering foils are compared with depth-dose curves of a 40-MeV linear accelerator (CGR, 1973) with field homogenized by scanning with the unscattered primary electron beam (Fig. 1), the advantage of the scanning technique, especially for energies above 20 MeV, can be seen at the higher depth doses and by the steeper decrease of the curves with depth. The handicap of the scattering-foil technique is mainly that the electrons lose energy in the foils of heavy metal. The energy loss is not uniform, so that after passing the scattering foil the energy spectrum is considerably broadened.

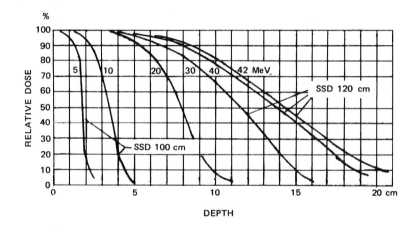

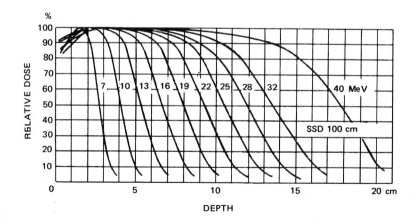

Figure 1. Depth-dose curves of electrons in water for (a) 42-MeV betatron with scattering foils for field homogenization (Schittenhelm et al., 1965) and (b) linear accelerator Sagittaire using scanning of the field with the unscattered pencil beam for field homogenization (CGR, 1973). Field sizes: (a) 12 cm diameter and (b) 10 cm × 10 cm.

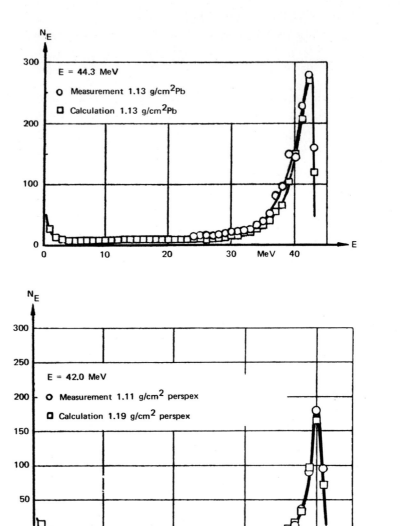

Figure 2. Energy spectra for 44.3-MeV and 42.0-MeV electrons measured after traverse of layers of 1.13 g/cm² and Pb and 1.11 g/cm² perspex (Reprinted, with permission from Benedetti, 1973).

In Figure 2 the transmission spectrum of 44-MeV electrons is presented after traversing (a) 1.13 g/cm² lead and (b) 1.1 g/cm² lucite (Benedetti, 1973). At field sizes up to 20 cm × 20 cm, a scattering foil-to-surface distance of 100 cm, and an electron energy of about 45 MeV, scattering foils of lead up to 1.3 g/cm² are needed. The mean energy of the shown lead-transmission spectrum is 5.8 MeV lower by radiation losses of the electrons

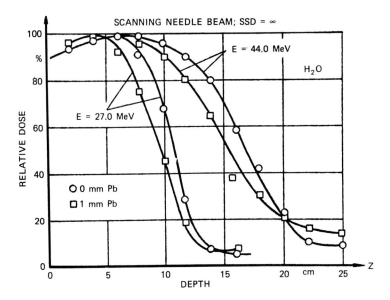

Figure 3. Depth-dose curves of 44.0-MeV and 27.0-MeV electrons in water with and without a scattering foil of 1.13 g/cm² Pb. Calculated for an infinite field size with field homogenization by scanning with a needle beam of primarily monoenergetic electrons. SSD=∞ (Benedetti, 1973).

than the most probable energy of 42.2 MeV. A total mean energy loss of 7.8 MeV results. This explains the relatively poor depth-dose curves of high-energy electrons when scattering foils are used. It can be shown that for larger field sizes and 100 cm scattering foil-to-surface distance, the mean electron energy after passing the scattering foil, even when the primary energy is increased, cannot be raised higher than about 40 MeV, because of the necessary increase of the foil thickness (Benedetti, 1973). Figure 3 demonstrates how the depth-dose curves of 27- and 44-MeV electrons are changed by a lead foil of 1-mm thickness.

The electrons do not come out of a betatron as a pencil beam as they do with a linear accelerator. Instead they are emitted as a diverging fan-shaped beam running within the accelerating plane. The dose in the irradiated field can be homogenized without a scattering foil by a pendulum movement of the fan beam or of the patient. With the so-called "telecentric, small-angle, pendulum technique," proposed by Rassow (1970), a converging irradiation of the tumor in directions parallel to the pendulum plane can be produced by a telecentric rotation of the betatron perpendicular to the acceleration plane (Fig. 4). For the 42-MeV Siemens betatron Rassow proposes a source-to-surface distance of 90 cm and a source-to-rotation-axis distance of 120 cm, i.e., a rotation axis at 30 cm below the body surface. The telecentric rotation provides the combined advantages of concentric irradiation and elimination of a scattering foil, so that acceptable dose distributions can be obtained with a relatively low surface-dose (about 70% for 40-MeV electrons) and with a dose maximum shifted into the depth. A set of depth-dose curves for rotation angles of 30° and greater is shown in Figure 5 (Rassow, 1972). Typical isodose distributions for

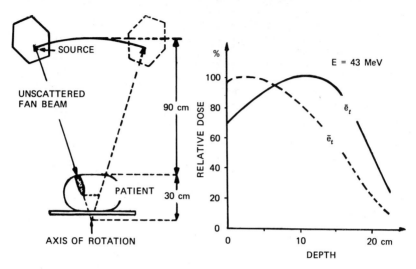

Figure 4. Telecentric rotation technique with the unscattered, fan-shaped, electron beam of a betatron. (a) Scheme of the technique and depth-dose curve for 43-MeV electrons in water (e^-_t). (b) For comparison, the depth-dose curve of a fixed beam of 43-MeV electrons with use of a scattering foil is also shown (e^-_f).

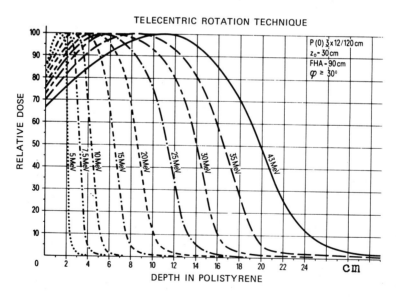

Figure 5. Depth-dose curves for telecentric rotation technique with electrons of various energies (Rassow, 1972). Rotation angle, $30°$ or greater; SSD, 90 cm; radius of rotation 120 cm, diaphragm opening corresponding to 3 cm × 12 cm at 120 cm SD (3 cm field width parallel to the pendulum plane which is perpendicular to the acceleration plane).

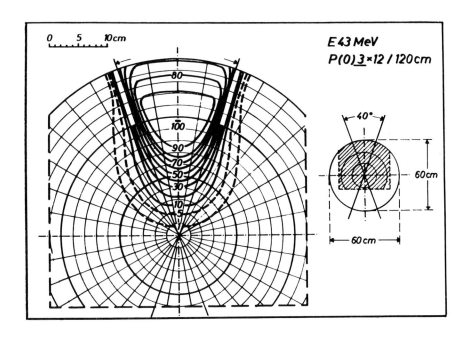
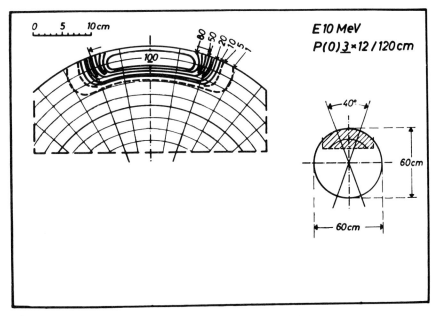

Figure 6. Isodose plans in the pendulum plane for telecentric rotation technique with 43- and 10-MeV electrons and a rotation angle of 40° (Reprinted, with permission from Rassow, 1970).

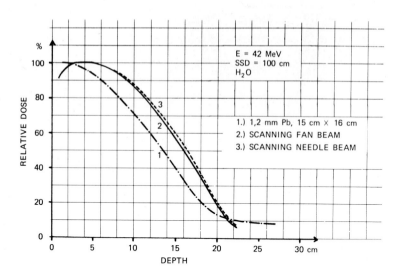

Figure 7. Comparison of depth-dose curves in water for divergent beams of 42-MeV electrons with different methods of field homogenization. SSD, 100 cm. Curve 1: Scattering foil of 1.2 mm Pb. Curve 2: Scanning fan beam. Curve 3: Scanning needle beam.

a rotation angle of 40° with 10- and 42-MeV electrons are shown in Figure 6 (Rassow, 1970).

Curve 3 in Figure 7 is a depth-dose curve for 42-MeV electrons when the scanning method with a needle beam for field homogenizing is used (Benedetti, doctoral dissertation, 1973). Curve 2 was theoretically calculated by transformation of a depth-curve for telecentric rotation with 42-MeV electrons to a depth-dose curve of a diverging radiation beam with the same source-to-skin distance of 100 cm as for curve 3, taking merely the geometrical relations as a basis. Curve 2 is practically identical with curve 3. This proves that one has the same energy spectral and the same relatively low integral doses with both kinds of field homogenization.

The scattering of the electrons in a scattering foil, however, does not seem to impair the dose distribution other than in energy loss. The dose distributions perpendicular to the beam axis are, at least at greater depths, nearly the same with scattering foil or with telecentric rotation, as shown in Figure 8 for 43-MeV electrons in a phantom depth of 11 cm.

Up to what depths of tumor and electron energy is there an advantage in using electrons instead of photons? In considering this question, one can use the areas under the depth-dose curves to measure the integral doses, if the same geometry of the radiation fields is given. In Figure 9 the depth-dose curves for ^{60}Co-γ-rays, 8- and 42-MV x-rays and for electrons of suitable energies (these are the heavy curves) are plotted, normalized at 5-, 10- and 15-cm depth of water. The source-to-skin distance is 100 cm, and the field size is 10 cm × 10 cm (Fehrentz, 1973).

For the choice of electron energies, the depth-dose curves for telecentric rotation have been taken as a basis. These are the heavy dotted lines in Figure 9. The electron energies

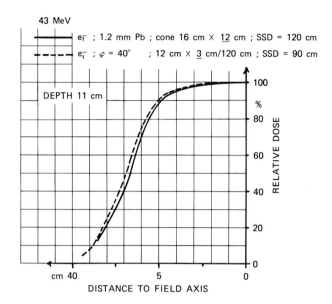

Figure 8. Dose distribution perpendicular to the field axis and to the accelerating plane of a 42-MeV betatron at 11-cm depth in a polystyrene phantom. Dotted line: Telecentric rotation of 40° rotation angle, 90 cm SSD and 120 cm rotation radius. Solid line: Scattering foil of 1.2 mm Pb, cone of 16 cm × 12 cm opening at 120 cm SSD.

have been chosen in such a way that a tumor of 10-cm diameter, lying with its center at the normalization depth of 5-, 10-, or 15-cm, still receives at its furthermost edge 80% of the dose in the center. For photon irradiations this is always the case.

The electron, depth-dose curves (heavy lines) have been mathematically obtained by the transformation of the depth-dose curves for telecentric rotation, as already explained, for a divergent beam with 100-cm source-to-surface distance. When these theoretical curves are normalized at the same depth, they represent the same integral dose and can therefore be compared with the photon depth-dose curves.

If we now compare the areas under the depth-dose curves in each of the three diagrams (Figs. 9a, b, c) up to a body thickness of twice the depth of normalization, i.e., 10, 20 and 30 cm, then in each case the areas under the electron curves are the smallest. Hence it would follow, the integral dose with electrons, using a field homogenization method without scattering foil, is smaller than the integral dose with photons for tumor depths up to about half the body thickness. Because there are body diameters up to 30 cm, at least in diagonal directions, there may be an advantage to using electrons in radiation therapy up to 40 or 50 MeV.

However, in the preceding considerations, the differences in the dose fall-off at the beam edges were neglected. With electrons you need a broader beam than with photons, and therefore integral doses for the necessary enlargements of the field sizes have been calculated (Fehrentz, 1973).

Figure 10 shows the integral doses of 8- and 42-MV x-rays and for electrons of suitable energy, compared with ^{60}Co-γ-rays, in relation to the depth of the tumor center. The

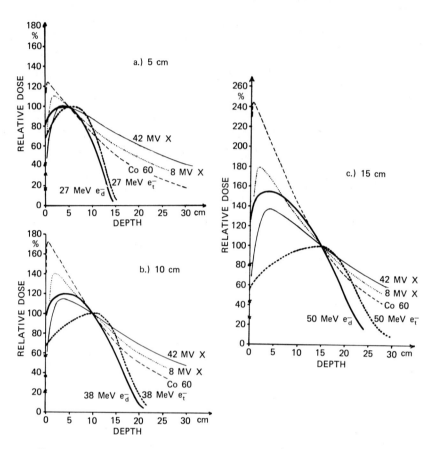

Figure 9. Depth-dose curves in water at 100 cm SSD for ^{60}Co-γ-rays, 8 and 42 MV x-rays, divergent electron rays of suitable energies using a scanning technique of field homogenization ($e^-{}_d$), and partly convergent electron rays with telecentric rotation technique ($e^-{}_t$). Normalization at 5-, 10- and 15-cm depth (Reprinted, with permission, from Fehrentz, 1973).

tumor diameter is 10 cm, and the body diameter 30 cm. Because of the need to widen for single- and multiple-field techniques, the different curves are given and designated S and M, accordingly. There are different curves for electrons with telecentric rotation and with scattering foil, and these are indicated by $e^-{}_t$ and $e^-{}_f$.

If we now, for instance, consider a tumor, lying with its center at a 10-cm depth, we have about 25% less integral dose with electrons given by telecentric rotation than with x-rays. Because the tumor in our example reaches up to 15-cm depth, we need an electron energy of 40 MeV. From these results (Fehrentz, 1973) one can say that for skin-to-tumor-center distances up to one third of the body diameter, electrons have an appreciably lower integral dose than photons.

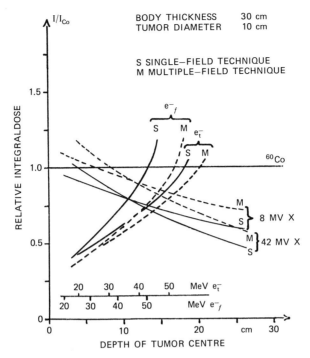

Figure 10. Integral doses related to ^{60}Co-γ-radiation as a function of depth of tumor center for single radiation fields of 8 and 42 MV x-rays and electrons of suitable energy using scattering foils (e^-_f) and telecentric rotation technique (e^-_t). S: Single field irradiation. M: Isocentric multiple field or rotation irradiation. Body thickness, 30 cm. Tumor diameter, 10 cm: SSD 100 cm except 90 cm for telecentric rotation with electrons (Reprinted, with permission, from Fehrentz, 1973).

CLINICAL EXAMPLES OF TELECENTRIC ROTATION WITH ELECTRONS

Telecentric shell irradiation with about 10-MeV electrons (Fig. 11) has the advantage of irradiating large areas without cold and hot spots. This method has already been described by Becker and Weitzel (1956). Particularly advantageous is telecentric rotation therapy of the bladder with electrons above 25 MeV (Fig. 12). The rectum and the superficial areas are spared very well. This technique is practiced in some centers and has also been used in our hospital for about 3 years. The irradiation is tolerated better than with ^{60}Co-γ-rays, and a higher survival rate seems to result. At least, this is the impression of the physicians who use it in our clinic.

Irradiation of the pelvic wall for gynecological residual tumors with telecentric rotation of about 35-MeV electrons is also used because of favorable dose distributions (Fig. 13). Another indication for using telecentric rotation therapy is in the treatment of a kidney with local lymphatic nodes. In Figure 14, 43-MeV electrons are used at a diagonal body thickness of 25 cm.

An interesting combination of electrons and photons to irradiate a kidney and the paraaortic lymph nodes has been proposed and is also practiced by Sack and Rassow

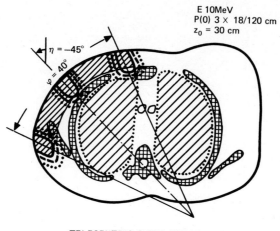

Figure 11. Scheme of telecentric shell irradiation of the chest wall with 10-MeV electrons, after Rassow (1972).

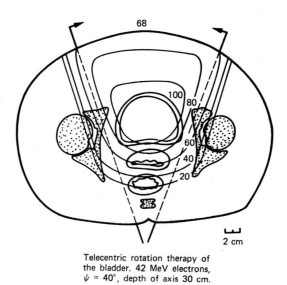

Figure 12. Isodose plan of telecentric rotation therapy of the bladder with 42-MeV electrons. Rotation angle, $40°$; depth of rotation axis below the body surface, 30 cm (Reprinted, with permission, from Poser et al., 1973).

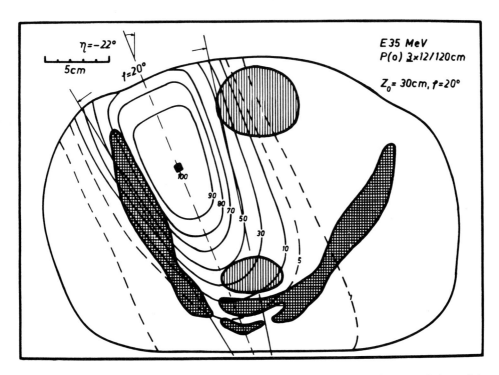

Figure 13. Isodose plan of telecentric, small-angle, pendulum irradiation of the pelvis wall for gynecological residual tumors with 35-MeV electrons. Rotation angle, 20° (Reprinted, with permission, from Rassow and Sack, 1971).

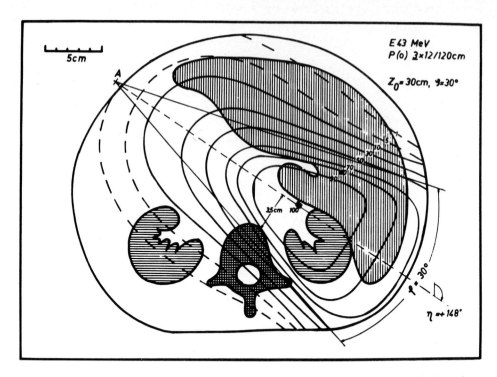

Figure 14. Isodose plan of telecentric, small-angle, pendulum irradiation of the kidney and the local paraaortic lymph nodes with 43-MeV electrons and 30° rotation angle (Reprinted, with permission, from Rassow and Sack, 1971).

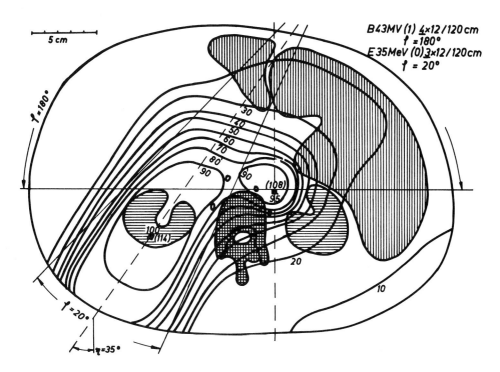

Figure 15. Irradiation of the kidney and the paraaortic lymph nodes by combination of a telecentric rotation of 20° with 35-MeV electrons and an isocentric rotation of 180° with 43 MV x-rays (Reprinted, with permission, from Sack and Rassow, 1972).

(1972). It is a telecentric rotation of the right kidney with 35-MeV electrons at an angle of 20° and an isocentric rotation of the paraaortic lymph nodes on the other side with 43-MV x-rays at an angle of 180° (Fig. 15). The second kidney, the intestinal area, the liver and the spinal cord are spared very well.

There are many other possible uses of electrons at energies above 25 MeV, especially in combination with photons. We believe electron therapy has brought an improvement in cancer management and will bring more with its further development.

REFERENCES

Becker, J. & G. Weitzel. 1956. Neue Formen der Bewegungsbestrahlung beim 15 MeV-Betatron der Siemens-Reiniger-Werke. *Strahlentherapie* 101:180–190.

Compagnie Générale de Radiologie (CGR). 1973. Prospectus of the Linear Accelerator "Sagittaire".

Fehrentz, D. 1973. Physical aspects of deep therapy with electrons and roentgen radiation up to 50 MeV. *Electromedica* 2:68–79.

Poser, H., G. Németh & H. Kuttig. 1973. Telezentrische Kleinwinkel-Pendelbestrahlung der Harnblase mit schnellen Elektronen. *Strahlentherapie* 145:390–395.

Rassow, J. 1970. Beitrag zur Elektronentiefentherapie mittels Pendelbestrahlung, IV. Mitteilung: Über eine neuartige, für primär unaufgestreute Elektronen spezifische telezentrische Kleinwinkel-pendeltechnik. *Strahlentherapie* 140:156–172.

Rassow, J. 1972. On the telecentric small-angle pendulum therapy with high electrons energies. *Electromedica* 1:1–5.

Rassow, J. & H. Sack. 1971. Beitrag zur Elektronentiefentherapie mittels Pendelbestrahlung, V. Mitteilung: Anwendung der telezentrischen Kleinwinkel-Pendeltechnik zur Strahlentherapie von Tumoren im Abdominalbereich. *Strahlentherapie* 141:5–12.

Sack, H. & J. Rassow. 1972. Beitrag zur Elektronentiefentherapie mittels Pendelbestrahlung, VI. Mitteilung: Strahlenbehandlung von Nierentumoren mit schnellen Elektronen und 42-MeV-Bremsstrahlen. Methodik und erste Erfahrungen. *Strahlentherapie* 144:641–648.

Schittenhelm, R., W. Derndinger, F. Groh, W. Gscheidlen, H. Haubold, F. Petersilka, P. Schipper, O. Steinmetz & R. Weiss. 1965. Ein Betatron für Elektronentiefentherapie. *Strahlentherapie* 127:578–628.

Discussion

Dr. Brady: Doctor Schumacher, would you like to make any comments?

Dr. Schumacher: Last year we had an x-ray congress in Berlin, and a one-year survival reported then would now be a two-year survival in the statistics reported here. Dr. Pohlit, how can we measure the hot spots; can we see them? In simulating our patient treatments with measurements in the Alderson phantom with film, we haven't seen these hot spots.

Dr. Pohlit: The data which I have shown were obtained in the following way. We would take a certain cross-section of the body with certain inhomogeneities, and, calculating from the data which I have shown, build a phantom and measure with very small detectors. (Very small detectors are necessary to avoid integration—if you use very large detectors, you integrate from the addition of dose to the suppression of dose, and everything is cancelled out.) We used an ionization chamber 1 mm in diameter and very thin rods of thermoluminescent material to measure the dose distributions. We also checked these as far as we could in patients.

Dr. A. Dutreix: We have also made similar measurements with film (on film you can see hot spots and cold spots very easily), and we have found hot and cold zones similar to those demonstrated by Dr. Pohlit.

Dr. Pohlit: We have actually measured, using real compact bone, and we found that the maximal dose goes down with increase of energy. The maximum dose increase was about 5% above 20 MeV, but only if the compact bone was about 2-cm thick. Thick compact bones are not found in the body, so the scattering due to bone has no bearing on the therapy.

Dr. Perez: How, when one is to use computers for treatment planning, can these inhomogeneities with electron beams be handled? What kind of computer programs are necessary, and what accuracy would you be able to develop?

Dr. Pohlit: We have known about hot spots for, I would guess, ten years, and at first we thought that one could deal with them only with Monte Carlo calculations. Since Monte Carlo calculations are very expensive and uncertain, we looked into the basic mechanism of these reactions. I think that we have arrived at simple assumptions which we can use. We do hope to be able to use a combined analog and digital computing method for measuring electron dose distribution, but we are not sure of the easiest way to do this. The state of the art at the moment is such that we have a general but not specific knowledge of how these problems can be handled. These data are still new—one year old—and we have not thought very much about applying them to the computer. But rather than say that there are tremendous hot spots and electrons shouldn't be used, I would like to say, we know a little more about both hot and cold spots. They are not mystic; they do exist. But it seems that we can deal with them in a quantitive way, and therefore they pose no danger.

Dr. Brady: Dr. Dutreix, Dr. Marbach asks: "Have you verified that electrons are, in fact, the cause of the observed shift in depth of dose maximum in your data?"

Dr. Dutreix: I am not sure I understand the question. I do not see any physical reason for a shift of the depth of the maximum due to photons. I think it certainly occurs with electrons. Of course, what we have done is to shield the beam at a distance from the phantoms, but we have not observed any difference in the depth of the maximum. The only difference we have seen is the modification of the spectrum due to shielding by a lead filter.

Dr. Brady: Dr. Pohlit, your data appear to assume a plane beam of electrons. If so, what effect would you expect from first, a divergent beam, and in the limit, an isotropic source?

Dr. Pohlit: The divergent beam would not change the results very much, at least if the patient is not treated very near to an accelerator. What is an isotropic source?

Dr. Marbach: Well, a source that isn't a point source or that isn't a plane beam. The beam of electrons, say, goes into the patient or the phantom, and becomes less of a plane and more isotropic. Your figures seem to indicate, or assume, that you had a plane beam of electrons.

Dr. Pohlit: Yes, the beam was homogeneous. Measured across, the beam would be constant within 5%.

Dr. Marbach: But as the beam moves into the phantom or into the patient, one would expect the parallelism to lessen. Would that affect the geometry of these hot spots?

Dr. Pohlit: No, it should not.

Dr. Brady: Dr. Pohlit, there is some concern about the absorbed dose in bone from electron beams. Do you have any numbers of comparison with high-energy photon beams?

Dr. Pohlit: Well, in electron beams, there is an increase of absorbed dose in the bone of about 10%.

Dr. Brady: Any other comments from this panel relating to that particular point?

Dr. Fehrentz: In the inner bone where there is soft tissue, there is no increase at all.

Dr. Pohlit: Bone marrow has a density of one, so there is no increase.

Dr. Perez: That's very important from a clinical viewpoint. When there is bone that's thick enough to perturb the dose beyond that point, that has to be taken into consideration by the clinician.

Dr. Brady: It can act almost as a shield to the underlying tumor.

Dr. Perez: Correct. And it can be significant, perhaps 10 to 20% with tonsillar lesions with energies below 18 MeV.

Dr. Brady: Dr. Ho raises a question relative to electrons at energies over 40 MeV. Are you concerned about the effects of neutron production on the staff who are setting up patients for treatment? That is, the induced radioactivity in the patient, the staff moving the patient, and the instrumentation in the room?

Dr. Pohlit: There is no activity in lead. But if there is copper, then the absorbed dose rate at a distance of 1 meter, is about 5 millirads per hour.

Dr. Fehrentz: I'd like to make a comment. The dose due to fast neutrons has been measured for a 35-MeV electron-beam from a betatron. The dose in the beam in the patient was lower than one per thousand. For the x-ray beam of 35 MeV, there was a dose contribution by fast neutrons of about 1%.

Dr. Ho: What worries me is the repetition of even this tiny dose several times a day, many days a year. What sort of an effect will this have on the staff? Especially the neutrons. All you need is a very small dose; it is cumulative.

Discussion

Dr. Pohlit: When handling the patient and going into the room with the machine shut down, you have no neutrons in the room.

Dr. Dutreix: Using film dosimetry, not only for photons but also for neutrons, we have checked the doses received by our technicians and they were always under the minimum detectable level.

Dr. Brady: Dr. Dutreix, you mentioned in your presentation that the penumbra of high-energy photon beams increases with the increase in field size.

Dr. Dutreix: The penumbra, that is the distance between the 90% and 10% isodose, increases with photon energy. It increases only slightly with the field size, but this is probably due to an increase of scattered photons.

Dr. Brady. Dr. Fehrentz, how were the telecentric dose distributions measured and determined for the high-energy electrons?

Dr. Fehrentz: We used film dosimetry, and an atlas of dose distributions for telecentric rotation therapy supplied by Siemens.

Dr. Brady: Was that used in individual calculations for each patient, or in treatment planning?

Dr. Fehrentz: Because there is not much change between specific individuals, we used standard dose distributions.

Dr. Paliwal: In the Varian accelerator a scattering foil is used, whereas in the Sagittaire accelerator there is a scanning mechanism. Dr. Pohlit has suggested that with the scanning mechanism, there is better depth-dose distribution. However, the data on the Varian accelerators are compatible, even though they use scattering foils.

Dr. Pohlit: Up to 20 MeV there is little difference between the two. At higher energies, the difference becomes important. I spoke of energies above 20 MeV for depth therapy.

Discussant: It seems that from your depth-dose data for electrons, the surface dose decreased with the increase in energy.

Dr. Pohlit: If you take higher energies, the dose maximum occurs deeper in the body. As in isocentric rotation therapy, the surface dose goes down.

Dr. Brady: Dr. Perez mentioned a 10% overdosage in bone for 20-MV photon beams. Dr. Dutreix thought that seemed high and wanted to know whether that is indeed what was meant.

Dr. Perez: We were talking about the absorption of energy. I'm going to refer that question to Dr. Purdy, who is a co-author of my presentation.

Dr. Purdy: The slide used in Dr. Perez' talk showed the absorbed dose in soft tissue within bone; for the 25 MeV accelerator it averaged about 20%.

Dr. Brady: That's not actually in bone itself, but in the soft tissues within bones.

Dr. Purdy: That's correct, in soft tissue the matrix is not really important.

Dr. Dutreix: Did you use the full spectrum of the 20 MV x-ray beam, or a 20 MeV monochromatic beam?

Dr. Purdy: What we used should be equivalent to about a 12- or 10-MeV monochromatic beam.

Dr. Brady: Dr. Perez, you mentioned a few cases of severe dental decay and osteonecrosis of the mandible in patients who were treated for carcinomas of the tonsils. Do you relate that to the bone dosage?

Dr. Perez: Others have been able to show a greater correlation between dose and osteonecrosis of the mandible—Fletcher several years ago, and Chang and Wang last year.

But here in our presentation only one of the cases really could be ascribed to a high dose, about 6900 rads, to the mandible. The other two patients had teeth problems, and there were tooth decay, cavities, periodontal disease, and subsequent myelitis, which was the initiating factor for the osteonecrosis.

Dr. Brady: What is your policy regarding teeth and the irradiated volume?

Dr. Perez: If the teeth are in bad condition, they are extracted before we start the therapy, and we wait at least two weeks for healing. If they are in relatively good condition, needing only minimal care, the dentist and hygienist take care of them before we initiate therapy. We do not extract good teeth. Our program of daily prophylactic fluoridation and periodontal care has significantly improved the subsequent state of the teeth and decreased the number of complications.

Dr. Brady: Recently there's been a suggestion that one might influence zerostomia by the addition of prostigmin derivatives during the course of the treatment in order to keep the salivary glands functioning. Do you think that's worthwhile or not?

Dr. Perez: I don't know.

Dr. Brady: Dr. Johns, do you want to comment on the question of mandibular osteonecrosis and severe tooth decay relative to high dosage?

Dr. Johns: I think the figure of 10 or 20% extra absorption in bones is a bit high.

Dr. Fehrentz: For 42 MV x-rays, we took real compactor and measured the dose in the compactor by ionization methods. The dose elevation was 8%. If you compare the depth-dose curve in water and the depth-dose curve with compactor in water, the dose elevation of 8% in the compactor was with 2 cm of bone. In the body you don't have much thick bone, so it has no importance for therapy.

Dr. Dutreix: I wonder if there is not a confusion between differences in attenuation in bone and soft tissue due to bone density, and the absorbed dose in different tissues. I think that the absorbed dose in a gram of bone is almost no different from the dose in a gram of soft tissue. But, of course, the attenuation is different, because the densities of bone and of soft tissue are different.

Dr. Perez: I think we are using the "absorption of energy" term properly in terms of the thickness of material rather than the individual density, and that is what is important in clinical therapy.

Dr. Brady: Dr. Perez, in your set-ups for treatment of carcinoma of the cervix, you have shaped the beam at the corner. Does that significantly influence the isodose distributions within the volume being irradiated?

Dr. Perez: We don't believe so. There is so little dependence on beam size with that energy that I would not expect any significant influence.

Dr. Brady: We've done measurements on that and there is no significant alteration of distribution. Here is another question: "Can you relate the treatment of carcinoma of of the larynx, in terms of the incidence of cartilaginous necrosis in the patients, to dose in terms of rads?" That would relate in some respect to the work done by Shukovsky.

Dr. Perez: We did have two cases of cartilaginous necrosis in about 60 patients with T_1, T_2 tumors of the larynx, and they were both related to a high dose—over 7000 rads.

Dr. Brady: This question is to the entire panel. It concerns the suggestion of integral dose, first put forth by Mayneord, and used at varying times in discussion of volume effects. What is the specific importance or validity of the concept for radiation therapy today?

Dr. Perez: This is an elusive concept to me, because I have never quite understood how the calculations are made, or how the dose can be integrated in this large volume. However, there is a difference between a large dose given to the entire volume with a certain energy, compared to another dose that has a greater penetration. Further, there is a relationship of volume to tolerance, which touches on integral dose concept, volume plus total dose.

Dr. Pohlit: Integral dose is not very difficult to calculate. We can calculate everything! But what does it mean? In general, the integral dose does have an effect. If you irradiate the whole body with an absorbed dose of 300 rads, the patient is dead. But, where the dose is very concentrated, it doesn't matter much if you concentrate it a little bit more or little less.

Dr. Perez: I don't quite agree with that. More complications occur with a relatively small gradient of dose within a volume. Let's consider pelvic irradiation, with two opposing AP and PA portals, as opposed to three or four portals, or rotational techniques. The patients that are treated with the more complex technique have a larger amount of small intestine being irradiated, and they have more gastrointestinal complications. The question is, what is the biological effect of the total dose to the patient? Whether it's transient or not, doesn't matter.

Dr. Johns: I like to look at it a little differently. No one believes in integral dose, I admit; it's just a way of talking about how much energy is absorbed in the patient per tumor dose. If that ratio is different from one treatment to another, and you put more radiation in the normal tissue, there is radiation in the tumor volume. It is a valid concept.

Dr. Brady: Might there be any relationship between the integral dose and the anticipation of damage to normal tissue?

Dr. Johns: Surely.

Dr. Perez: Yes. We've been very concerned with measured complications and real morbidity. But we've got to begin worrying about subclinical damage. There's a lot of injury we cause that we don't let bother us, yet the patients are having problems. Their tissues should be spared as much as possible. The quality of life is very important.

Dr. Kramer: Considering dose absorbed in normal tissue versus the tumor, we are facing the same trade-off of either giving a little dose to a lot of normal tissue, or a lot of dose to little normal tissue. I very much question that there is more absorption in the tissue because of the technique used. I think it depends on the particular beam used, but for the same dose in the tumor, a rotational technique does not necessarily give more absorption in normal tissue. You have to establish what kind of dose you are talking about and for what kind of normal tissue. In some tissues you can afford to give a modest dose and not get into trouble, and in other tissues it's better to give a high dose to little tissue.

Dr. Perez: Small doses to the gastrointestinal tract may cause no problem. But we have never studied intestinal absorption after those doses. With lung tissue, you can give small doses and nothing happens. But let's do pulmonary function studies and see if they have had a 5 or 10% decrease in capacity. Those are the little intangibles that we don't worry about; nevertheless, they are there.

Dr. Powers: We will have to specify our dose in other terms if we want a given dose at a given point. Profiles of dose, comparing and contrasting cobalt with betatron beams, are a tool for a physician to construct in his own mind a comparison of two treatment techniques. I am not really able to subtract one isodose curve from another and consider the difference; I have considerable difficulty telling which is appropriate. What we really need is a profile of tumors, the number of cells, and the dose needed, with a gradient including the appropriate dose to the tumor. The shrinking field offers a variety of doses, because of the variety of loading. We should develop a profile that includes the tumor in the dose desired, and optimally reduces the dose in the normal tissue. We are able to develop techniques, and it is necessary to do so. The higher-energy devices have permitted me to actually reduce the entire volume of the dose. There are tissues that will tolerate a very low dose; and they are not important. But in both the transient tissue and the interstitial tissue, there are going to be limiting characteristics. It's an area that needs to be explored by both clinicians and other individuals concerned with high-LET radiations.

Dr. Brady: "Is the integral dose being tabulated in any of the current national clinical trials?" I think not. "Dr. Johns, in doing the calculations for the integral dose, are there mechanisms to compensate for different machine modalities and problems relating to dose distribution?"

Dr. Johns: That calculation is based on just one distribution. Given a 10 cm X 10 cm field for 25 MeV and a tumor at a depth of 10 cm, you find that the integral dose given to the patient for a given tumor dose is higher for 200 kV than for cobalt, which, in turn, is higher than for 25 MeV, in turn higher than for 35 MeV, leaving the field size constant.

Dr. Brady: "Dr. Schumacher, when you speak of the 6000-rad dose given at the rate of 500 rads weekly for a period of 12 weeks, is that the tumor dose or, if not, where is that measured?"

Dr. Schumacher: That is the tumor dose, in each case first measured by the Alderson phantom. When I say 500 rads that means the maximum dose; about 80% of the maximum dose is a tumor dose. Therefore, in each week we give about 400 rads to the tumor.

Dr. Brady: That is, 400 rads per fraction once a week for 12 weeks?

Dr. Schumacher: We are measuring this from the Alderson phantom, so we don't need any correction data.

Dr. Brady: Are you correcting for lung inhomogeneities?

Dr. Schumacher: Yes.

Dr. Simpson: There seems to be some confusion. The machine is not giving you anything at all. It all depends on how somebody is calculating from the machine, and electron beams can be calibrated in terms of rads, which is absorbed dose. What Dr. Schumacher is saying, is that the given dose, which is the dose at maximum, is 500 rads per treatment. So if you want to call it 80%, it's 400 rads per treatment for tumor.

Dr. Brady: With the fractionation schedule that you're using and the protraction program, what is the dose to the spinal cord. Is that of any concern to you?

Dr. Schumacher: We give a direct field ventrally and an oblique field dorsally, and so we do not irradiate through the spinal cord. The dose on the spinal cord is about 30%; we have about 1000 rads on the spinal cord. In all 5000 cases, we have never seen any complications from the spinal cord.

Dr. Brady: "What is the influence of the electron beam on pulmonary function studies? You quoted in your presentation the influence on vital capacity. Was that the only parameter of pulmonary function measured?"

Dr. Schumacher: No. The clinicians did all function studies, including oxygenation of the blood. We have never seen any difficulty in lung function studies when we altered our frequency.

Dr. Brady: You said the vital capacity actually improved.

Dr. Schumacher: We studied the lung function beforehand, and in cases of poor lung function, which posed a risk for treatment, we did not treat, except if the poor lung function was caused by compression from the tumor.

Dr. Roswit: Did you have a single case of the acute syndrome or any deaths from it in the 5000 patients? I would find it very hard to believe that you did not have such problems.

Dr. Schumacher: I showed you my statistics. Sometimes when we are treating, fibrosis develops.

Dr. Roswit: No deaths from the acute syndrome?

Dr. Schumacher: Maybe there are some, but we have none in our statistics.

Dr. Roswit: We used to call it bronchopneumonia. Maybe you don't recognize it because it may come so late; one month, two months or three months later, the patient dies. You may not even know about it or fail to recognize it as an acute radiation effect.

Dr. Schumacher: All our patients are controlled. Maybe if the patient were ambulant, it would be possible.

Dr. Brady: You made a very subtle implication a moment ago that the individuals who come to the lung clinic in Berlin are worked up from the pulmonary function point of view, and some are excluded if their functions are compromised?

Dr. Schumacher: Yes, only a few.

Dr. Brady: So that you're only irradiating those that don't have major compromise in their pulmonary functions. You are excluding some patients. How many are those? What percentage?

Dr. Schumacher: It's about 1%, not more.

Dr. Brady: Not many. Do the changes in the pulmonary function, following the completion of radiation therapy, continue to be improved, or do they revert to essentially what they were before the treatment program? In some studies in the United States on patients with carcinoma of the lung, most of the profiles of pulmonary functions are done without differential bronchospirometry, that is, on the total lung capacity and not on one side versus the other side. Some individuals have compromised function, but not to a degree that would prevent them undergoing a program of treatment. Subsequent to the completion of the treatment, they will manifest, for a period of time, transient improvement in their function. This relates to many different factors: control of the trachial bronchitis and of chronic pulmonary infection, relief of the obstruction of the bronchus as a consequence of the radiotherapy, and cessation of smoking (which, of course, you are influencing considerably in these patients because that is one of the first things you tell them to do). But then after about 1 to 3 months, the pulmonary functions have reverted to essentially what they were before the radiotherapy was initiated. Now, in your program of management, you were doing pulmonary functions on a continuing basis, I presume. Do you see the same pattern developing in the patients you're treating?

Dr. Schumacher: In some cases we do.

Dr. Brady: And what percentage of individuals maintain their improved function after completion of their treatment?

Dr. Schumacher: About 20 to 30%.

Dr. Brady: In our experience vital capacity is the least reliable factor; the more important ones are diffusing capacity, flow rates and oxygen and CO_2 measurements. Even though vital capacity is a relatively simple thing to measure, it's markedly influenced by various other factors totally unrelated to lung disease.

Can you explain why your end results are better than those reported in the general literature? Are there factors of selection? Are there other factors, such as the distribution of stages, that might help to explain why your patients do so much better than most of those reported in the world?

Dr. Schumacher: The most important factor is the use of high-energy electrons for concentration of dose in the tumor; the next factor is a lower total treatment time, and a better tolerating effect as a result. We started with 1000 rads per week, in palliative treatment, giving 1000 or 1200 rads per week. In some cases we gave 2000 rads per week to find out the tolerating effect of this high weekly dose. We reduced the initial high doses to 900 rads. We had also given 500 rads two times per week up to 3000 rads, and 400 rads two times per week up to 3000, and 500 rads once a week. We treated not only lung tumors with this scheme, but all our patients. We have seen from the 500 rads two times per week experience that you don't need this high a dose; the tolerating effect with 500 rads twice a week, is very bad. So we have changed all cases, going down to 500 rads once a week. With that treatment scheme there is a much better tolerating effect. I have heard of reactions with 4000 rads or 5000 rads in four weeks. Dr. Dutreix said he altered the treatment scheme in that he changed to single doses separated by one week, but he didn't change his treatment scheme to give 500 rads once a week. Maybe you can give 250 rads two times weekly, but if you reduce the weekly dose, you have a better tolerating effect. You can give more tumor dose if the tumor is not fully shrinking. That works the same with all the tumors. You can give 7000 rads with a good tolerating effect; that is the important thing. Rather than alter the frequency in six weeks, you could reduce the single weekly dose and then alter and get a better tolerating effect. There would not be so many complications if, instead of a weekly dose of 900 rads, you would lower it to 600 rads. The only reason to give a higher dose in a shorter time is when the tumors are growing, the tumor doubling rate being high. But in our patients, the median age is about 40 to 60 years, so the tumor doubling rate is very low, and you have time enough to treat.

Dr. Perez: We don't have so many complications; those are more side effects than anything else. Dr. Schumacher, have these patients you have treated received any systemic therapy—adjuvant chemotherapy or immunotherapy? Have you had any special supportive care program? Your figures are very impressive. The majority of patients with bronchogenic cancer fail, not because of local disease, but because of distant dissemination. I do expect a certain amount of improvement in survival by an aggressive local and regional approach, but not to the extent that you have shown us. I think your figures are remarkable. There must be some other factor. Maybe it's the air in Berlin.

Dr. Roswit: There may be a factor there. You have a remarkably high number of patients with well-differentiated, squamous-cell carcinoma. We do not expect that in patients, except in lesions that remain fairly local. We have 4000 autopsies in the VA program that now covers 8000 patients. The figures show that, of the cases of oat-cell carcinoma, 90% have disseminated disease; of adenocarcinomas, 86%. But of the squamous cancers, only 56% have disseminated disease. Now there is the focus for our attack. Maybe with the high percentage of squamous cancers you have, you're getting better control than we are. I can't see any other reason.

Section III

New Horizons for High-Energy Beams

The Impact of Radiosensitizers of Hypoxic Cells

Jack F. Fowler, D.Sc., Ph.D., F.Inst.P.,
Gray Laboratory,
Mount Vernon Hospital,
Northwood, Middlesex, England

G. E. Adams, D.Sc., Ph.D.,
Gray Laboratory
Mount Vernon Hospital
Northwood, Middlesex, England

Radiation sensitizers which act specifically on hypoxic cells have recently given remarkable results on experimental tumors in animals, and pilot studies on patients have started. These sensitizers should enable low-linear energy transfer (LET) radiation to deal with the problem of hypoxic cells in tumors. It is still not known whether this problem is an important clinical one, and if so, in what types of tumor. The hypoxic-cell sensitizers offer a way of answering these long-standing questions. It will be some years before it is possible to assess whether these sensitizers will cope clinically with hypoxic cells instead of, or perhaps together with, high-LET radiation.

Several different classes of radiosensitizing drugs have been described (Adams, 1967, 1973; Bridges, 1969; Emmerson, 1972). However, if radiation sensitizers are to be of any practical clinical value, they must cause more damage to tumors than to normal tissues. The largest difference in radiosensitivity known between tumors and normal tissues is due to the radioresistance of hypoxic cells in tumors. The presence of such cells has been demonstrated repeatedly in animal tumors and has been shown to limit the radiocurability of the tumors. This disadvantage of hypoxic cells is smaller in tumors which reoxygenate their hypoxic cells during radiation therapy.

Methods suggested to overcome the problem of hypoxic cells in tumors include treatment in hyperbaric oxygen chambers and radiation therapy with high-LET particle beams (Fig. 1). The third major method of overcoming the problem is the application of chemical radiosensitizers which are active against hypoxic cells only. Most sensitizers of this type are electron-affinic compounds and mimic the radiosensitizing effect of oxygen.

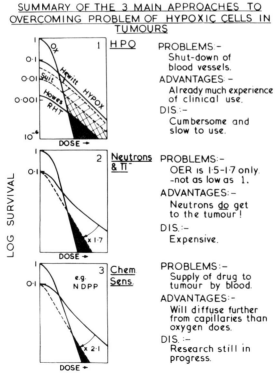

Figure 1. The three main methods of dealing with hypoxic cells in tumors, expressed as cell survival curve diagrams. The extent of the problem remaining after a single-dose treatment is indicated by the black area.

They are not used up in the metabolism of the cells through which they diffuse, so that they can penetrate further from capillary vessels than oxygen does.

In 1963, Adams and Dewey proposed that a relationship existed between the ability of a few chemical compounds to sensitize hypoxic bacterial cells and the electron affinity of the compounds. Subsequent work with bacterial systems (Ashwood-Smith et al., 1967; Adams & Cooke, 1969) and later with mammalian cells *in vitro*, led to the characterization of a large number of radiosensitizers. The electron affinity hypothesis was verified in the main, and thus aided the search for more active compounds. A report of the radiosensitizing properties of several nitrofurans, including some known urinary disinfectants, on hypoxic mammalian cells *in vitro* was significant because toxicological and pharmacological information was available for these compounds (Chapman et al., 1972). However, subsequent attempts to demonstrate appreciable sensitization *in vivo* with the nitrofurans have been disappointing, in some cases because the metabolic half-life was only a few minutes.

A significant step forward occurred with the successful demonstration of the sensitization, *in vivo*, of epidermal cells in mice that were made artificially hypoxic by breathing nitrogen for 35 seconds, the irradiation by electron beam being delivered between the 25th and 35th second (Denekamp & Michael, 1972). The compound used was NDPP, a soluble derivative of paranitroacetophenome which was known to be an active sensitizer *in vitro* (Adams et al., 1972). A small degree of radiosensitization was also observed in mouse ascites tumor cells irradiated *in vivo* (Berry & Asquith, 1974) and in solid tumors in mice (Sheldon & Smith, 1975). However, encouraging as these results were, the chemical instability of NDPP and its high attachment to serum proteins ultimately limited its usefulness (Whitmore, private communication).

Further searches for other drugs already in clinical use and possessing a chemical structure associated with electron affinity led to the discovery in 1973 of the radiosensitizing action of metronidazole (Flagyl, May and Baker Ltd., Dagenham, Essex, England) (Foster & Willson, 1973; Chapman et al., 1973; Asquith et al., 1974a). Although this sensitizer is only moderately active on a concentration basis, experiments in various types of systems *in vivo*, including solid tumors in mice, gave promising results. This was due to its low toxicity, wide distribution in tissues, and very importantly, its long metabolic half-life, all of which are properties necessary for a clinically useful hypoxic cell radiosensitizer.

Several other compounds related to metronidazole, a 5-substituted nitroimidazole, were investigated in attempts to find more active compounds. On theoretical grounds related to the effect of chemical structure on the electron affinity of the nitroimidazoles, it was anticipated that the 2-substituted nitro compounds might be more effective than the 5-nitro derivatives. Recently, one such compound has been shown to be even more effective than metronidazole (Asquith et al., 1974b).The current status of research on radiosensitization by this compound, Ro-07-0582 (Roche Products, Ltd., Welwyn Garden City, Herts., England) is summarized below.

SENSITIZATION BY NITROIMIDAZOLES *IN VITRO*

An investigation of the enhancement ratio for sensitization of x-irradiated Chinese hamster cells $V79$-GL_1 by both metronidazole and Ro-07-0582 showed that radiosensitization occurred within seconds of mixing the cells with medium containing the drug; that sensi-

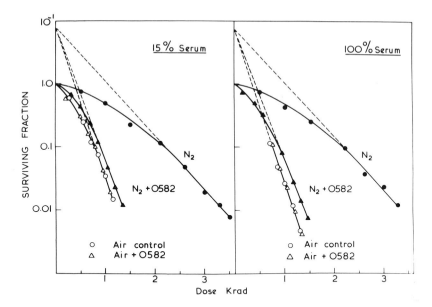

Fig. 2. In vitro survival curves for hypoxic and oxic Chinese hamster cells V79CL₁ irradiated in the presence of 4 mM Ro-07-0582. Sensitization is not diminished by the presence of serum in the culture medium (Asquith et al., 1974b).

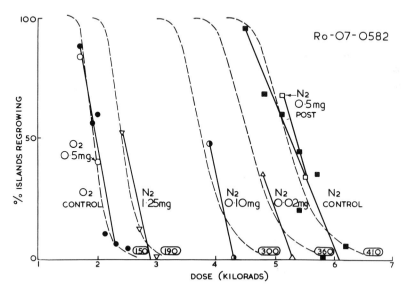

Fig. 3. Survival of epidermal clones (i.e., basal cells) as a function of x-ray dose for various doses of drug in milligrams per gram bodyweight. The animals were irradiated breathing nitrogen except for the left-hand curve when they breathed oxygen. The solid lines are "by-eye" fits. The dashed lines are computer fits with the calculated D_0 at the foot of each curve (Denekamp et al., 1974).

tization occurred at much lower concentrations than were cytotoxic after two hours' contact; and that for 0582 the maximum enhancement ratio was 2.5, i.e., nearly equivalent to the full oxygen enhancement ratio of 2.7 for this cell line. For both drugs it was shown that the sensitization efficiency was independent of the serum concentration in the medium (Fig. 2) and unaffected by the position of the cells in the mitotic cycle (Asquith et al., 1974a and b).

Other investigations have shown a specific cytotoxicity for hypoxic cells if they are exposed to the drugs for periods of many hours (Hall & Roizin-Towle, 1975). This specific killing of hypoxic cells should be a useful contribution to therapy, but in animal tumor experiments it contributed only a small part of the much larger effect of killing by radiation, for single-dose administrations.

SENSITIZATION BY NITROIMIDAZOLES *IN VIVO*

A variety of different assay systems has been used to test electron affinic sensitizers *in vivo*. Results for the two drugs of present clinical interest will be summarized here.

The first *in vivo* test system used was the Withers (1967) skin clone methods of assessing epidermal cell survival, in the skin of mice made briefly hypoxic for the present tests (Denekamp et al., 1974). Enhancement ratios of 2.1 and 1.5, respectively, for 0582 and metronidazole (1 mg/g body weight) were obtained, as compared with the oxygen enhancement ratio of 2.7 in this system. Figure 3 shows that no enhancement of cell killing in oxygenated cells was caused (left-hand curve) and that no sensitizing effect was observed when the drug was given after irradiation (right-hand curve).

A small degree of sensitization of skin reactions has been observed when large single doses of x-rays were used, over 2000 rads, due to the presence of a few hypoxic cells in normal skin, but no significant sensitization for three or five fractions of x-rays (1200–800 rads per fraction) was observed (Foster, 1975). Nevertheless, we may expect significant enhancement of radiosensitivity in any normal tissues which are hypoxic, such as cartilage.

SENSITIZATION BY NITROIMIDAZOLES OF SOLID TUMORS IN EXPERIMENTAL MICE

Promising results have also been obtained in various animal tumor systems investigated by workers at the Gray Laboratory in England, at the Ontario Cancer Research Institute, Toronto, Canada (Rauth, 1975), and recently in other laboratories, too. Table 1 summarizes these data for single doses of x-rays. It is clear that many types of tumor, tested in several different ways, give dose enhancement ratios greater than 1.7, which is the corresponding therapeutic gain factor for neutrons or negative pi mesons. Enhancement ratios are also shown in the table for the lower drug concentrations of 0.2 to 0.3 mg/g body weight, which corresponds in mice to the doses which can be administered to patients in practice.

The results for local control of tumors, i.e., cure of mice, are particularly encouraging. Figure 4 shows results for first-generation transplants of spontaneous mammary tumors in C3H/He mice (Sheldon et al., 1974). The tumors were irradiated when they

Table 1. ENHANCEMENT RATIOS FOR SOLID MURINE TUMORS *IN VIVO*

Type	Tumors Doubling Time (days)	Hypoxic Cells (%)	Assay	X-Ray Dose Enhancement With RO-07-0582 (0.2–0.3 mg/g)	(1mg/g)	Experimenter
						Gray Laboratory
CBA Fast Sarcoma F	1	10	Regrowth Delay	1.0	1.5	Begg
CBA Carcinoma NT	3	6	Loss of 125IUdR	1.4	2.2	Denekamp & Harris
WHT Bone Sarcoma	2.5	—	Regrowth Delay	—	1.8	Denekamp & Stewart
WHT Fibrosarcoma	2	—	Regrowth Delay	—	1.8	Denekamp & Stewart
CBA Fast Sarcoma F	1	18	Cell Dilution *In Vitro*	1.3	2.2	McNally
WHT Squamous Carcinoma D			Cell Dilution *In Vivo*		1.0 (i.v.)	Hewitt
WHT Intradermal Squamous Carcinoma G	1	0.3	Cure	1.9	2.1	Peters
C₃H 1st Generation Transplant of Spontaneous Mammary Carcinoma	6	10	Cure	1.7	1.8	Sheldon & Fowler
WHT Anaplastic MT Line Transplant	1	50	Cure	1.7	2.0	Sheldon
WHT Anaplastic MT	1	50	Cell Dilution *In Vitro*	—	1.5	McNally
						Other Laboratories
			Cell Dilution			Rauth (Toronto)
C₃H Sarcoma KHT	2	6	Lung Colonies	1.2–1.3	1.8	Kedar, Watson and
			Cell Dilution			
EMT 6			*In Vitro*	—	2.2	Bleehen* (London)
EMT 6			Cell Dilution *In Vitro*	—	2.4	Brown** (Stanford)
C₃H 3rd Generation Transplant of Spontaneous Mammary Carcinoma			Cure	—	2.4	Stone & Withers** (Houston)
C₃H Mammary Carcinoma			Cure	—	2.3	Brown** (Stanford)

*Private Communication   **Papers presented at the Annual Meeting of the Radiation Research Society, Miami Beach, Florida, May, 1975

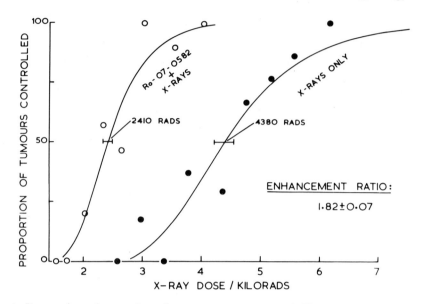

Figure 4. Proportion of transplanted mammary tumors in C_3H mice controlled at 150 days versus x-ray dose. Right-hand curve, x-rays only. Left-hand curve, x-rays delivered starting 30 minutes after i.p. injection of 1 mg/g bodyweight of Ro-07-0582 (Sheldon et al., 1974).

reached a mean diameter of 6.5± 1 mm, the mice being anesthetized with sodium pentobarbitone. A dose of 1 mg/g body weight of Ro-07-0582 was administered intraperitoneally (i.p.) 30 minutes before irradiation with 240 kV x-rays at 240 rad/min. Although the enhancement ratio of 1.8 is modest compared with other results in the table, the use of the drug increases the local control rate from 10% to about 90% at 3200 rads. Further, the dose-response curve was significantly steeper, in the same ratio of 1.8, with the 0582. This finding suggests that the drug was indeed reaching all of the hypoxic cells and sensitizing them efficiently. A similar conclusion follows from the results of Denekamp and Harris (1975) on regrowth delay in the carcinoma "NT" in CBA mice.

FRACTIONATED X-RAYS PLUS RO-07-0582

The mammary tumor system in C_3H mice has been used for extensive investigations on optimum fractionation with x-rays and neutrons (Fowler et al., 1972; 1974; 1975). Three of the fractionated schedules were repeated with and without Ro-07-0582 (0.67 mg/g i.p. 30 minutes before each irradiation). All three were fairly short schedules, of 4 or 9 days' overall time, in which the poor or mediocre results were expected to be due to the presence of hypoxic cells and their inadequate reoxygenation.

For the x-ray schedule in which the response was poor, a substantial gain factor of 1.3 (x-ray dose enhancement ratio) was obtained. For the two mediocre schedules, smaller enhancement ratios of 1.1–1.2 were found (Fig. 5). These enhancement ratios were less than the value of 1.8 found for single doses because reoxygenation had eliminated some

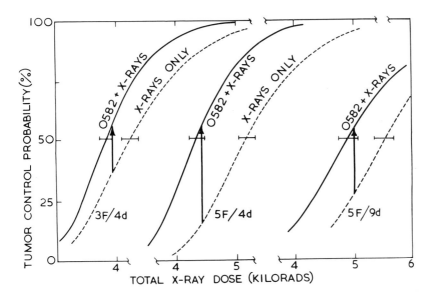

Figure 5. Proportion of transplanted mammary tumors controlled in C_3H mice for three fractionated x-ray schedules with and without the hypoxic-cell radiosensitizer Ro-07-0582 (0.67) mg/g. The horizontal error bars indicate the standard error of the mean $TCD_{50}s$. The vertical arrows show the improvement in tumor control at x-ray doses corresponding to a skin reaction of 2.0 (Fowler et al., 1976).

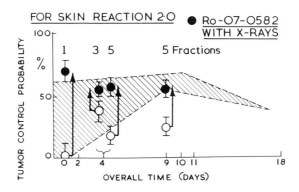

Figure 6. Tumor control probabilities (for a skin reaction of 2.0 from Fig. 5) versus overall time. Solid circles: x-rays only for the same schedules. The vertical arrows indicate the improvement in tumor control with respect to concurrent x-ray-only experiments. T symbols represent s.e.m. The shaded region represents the lines joining other fractionated x-ray schedules using 2, 5, 9 or 15 fractions (Fowler et al., 1976).

of the hypoxic cells in the fractionated schedules, and there was less disadvantage to gain back.

Figure 6 shows these results plotted for doses which give an average acute skin reaction of 2.0. The vertical arrows show the improvement in local tumor control obtained by the use of Ro-07-0582. It is impressive that all four schedules, including the single dose, bring the tumor control up to the same level of about 55–65%. This suggests that these sensitizers, like neutrons (Fowler et al., 1972), take the variability out of fractionated x-ray schedules. They would be particularly useful for non-standard schedules with fewer and larger fractions, because these appear to give more variable results than schedules with many small fractions.

PROSPECTS FOR CLINICAL APPLICATION

Although basic radiochemical and *in vitro* research has been in progress for some years, it is only within the last two years that substantial sensitization *in vivo* has been obtained, and progress is now rapid. Preliminary clinical investigations are in progress at Edmonton, Alberta, Canada (Urtasun et al., 1975) and at Mount Vernon Hospital, London, England (Dische et al., 1975). At Edmonton, several dozen patients with brain tumors have been treated by multiple x-ray doses with between 3 and about 10 g metronidazole given orally before each dose. At Mount Vernon, about a dozen patients have been treated with single or repeated doses of metronidazole, and eight advanced cancer patients have been treated with Ro-07-0582 (Dische et al., 1975). These patients had either multiple skin nodules secondary to breast or lung cancer, or nodules in the lung observable on x-ray films. The early conclusions are that about 10–12 g of metronidazole or about half that quantity of Ro-07-0582 can be administered orally and that the serum concentrations obtained correspond to enhancement ratios which were significant in mouse tumors (e.g., 165 μg/ml after 200 mg/kg metronidazole and 150–200 μg/ml after 100–150 mg/kg of Ro-07-0852). The tolerance limit of both drugs is set by nausea. Considerably higher concentrations have to be given, at more frequent intervals and continuing for longer periods, to cause the long-term brain damage reported in dogs (Scharer, 1972) for some nitroimidazole compounds, including metronidazole. Toxicological and pharmacological studies have been carried out by Johnson and colleagues at Roswell Park Memorial Institute, Buffalo, New York, and by Roche Products Ltd. in England.

Of four patients whose tumors yielded results, three demonstrated enhanced regrowth delay after single doses of x-rays with Ro-07-0582 (Dische et al., 1975). If such results are confirmed by further work, this drug will provide definitive evidence of whether hypoxic cells matter in radiotherapy. Research is proceeding in a search for radiosensitizers which will give equally good enhancement ratios as those demonstrated by Ro-07-0582, but for smaller drug doses and with no greater toxicity.

ACKNOWLEDGMENTS

The authors have pleasure in acknowledging the generous help and information given in the preparation of this review by his colleagues, especially Drs. Denekamp, McNally, Watts, Rauth, Bleehen, Urtasun, Dische, Strickland, Alan Gray, Wingate, Zanelli, Thomlinson, and Messrs. Foster, Sheldon and Begg. I should also like to thank the authors (as acknowledged) and the editors of the following journals for permission to reproduce figures: Fig. 1, *Excerpta Medica* (Proc. Int. Congr. Radiol. Madrid, Oct., 1973); Figs. 2 and 3, *Radiation Research*; Fig. 4, *British Journal of Cancer*; Figs. 5 and 6, *International Journal of Radiation Oncology, Biology & Physics*.

REFERENCES

Adams, G. E. 1967. The general application of pulse radiolysis to current problems in radiobiology. p. 35–94. *In* Ebert, M. & A. Howard (eds.) Current Topics in Radiation Research. Vol 3. North-Holland Pub. Co., Amsterdam.

Adams, G.E. 1973. Chemical radiosensitization of hypoxic cells. *Brit Med Bull* 29: 48–53.

Adams, G.E., J.C. Asquith, M.E. Watts & C.E. Smithen. 1972. Radiosensitization of hypoxic cells in vitro: a water-soluble derivative of paranitroacetophenone. *Nature New Biol* 239:23–24.

Adams, G.E. & M.S. Cooke. 1969. Electron-affinic sensitization: 1. A structural basis for chemical radiosensitizers in bacteria. *Int J Radiat Biol* 15:457–471.

Adams, G.E. & D. L. Dewey. 1963. Hydrated electrons and radiobiological sensitization. *Biochem Biophys Res Commun* 12:473–477.

Ashwood-Smith, M.J., D.M. Robinson, J.M. Barnes & B.A. Bridges. 1967. Radiosensitization of bacterial and mammalian cells by substituted glyoxals. *Nature (London)* 216:137–139.

Asquith, J.C., J.L. Foster, R.L. Willson, R. Ings & J.A. McFadzean. 1974a. Metronidazole ('Flagyl'), a radiosensitizer of hypoxic cells. *Brit J Radiol* 47:474–481.

Asquith, J.C., M.E. Watts, K. Patel, C.E. Smithen & G.E. Adams. 1974b. Electron-affinic sensitization: V. Radiosensitization of hypoxic bacteria and mammalian cells in vitro by some nitroimidazoles and nitropyrazoles. *Radiat Res* 60:108–118.

Berry, R.J. & J.C. Asquith. 1974. Cell cycle-dependent and hypoxic radiosensitizers. p. 25–36. *In* Advances in Chemical Radiosensitization. Proc. IAEA/WHO meeting on Radiation Sensitization and Protection. IAEA, Vienna.

Bridges, B.A. 1969. Sensitization of organisms to radiation by sulphydyl-binding agents. p. 123–176, *In* Augenstein, L.G., R. Mason & M. Zelle (eds.) Advances in Radiation Biology. Vol 3. Academic Press, London and New York.

Chapman, J.D., A.P. Reuvers, J. Borsa, A. Petkau & D.R. McCalla. 1972. Nitrofurans as radiosensitizers of hypoxic mammalian cells. *Cancer Res* 32:2616–2624.

Chapman, J.D., A.P. Reuvers, & J. Borsa. 1973. Effectiveness of nitrofuran derivatives in sensitizing hypoxic mammalian cells to x-rays. *Brit J Radiol* 46:623–630.

Denekamp, J. & B.D. Michael. 1972. Preferential sensitization of hypoxic cells to radiation in vivo. *Nature New Biol* 239:21–23.

Denekamp, J., B.D. Michael & S.R. Harris. 1974. Hypoxic cell radiosensitizers: Comparative tests of some electron-affinic compounds using epidermal cell survival in vivo. *Radiat Res* 60:119–132.

Denekamp, J.. & S.R. Harris. 1975. Tests of two electron-affinic radiosensitizers in vivo using regrowth of an experimental carcinoma. *Radiat Res* 61:191–203.

Dische, S., A.J. Gray, G.D. Zanelli, R.H. Thomlinson, G.E. Adams, I.R. Flockhart & J.L. Foster. 1975. Clinical testing of the radiosensitizer Ro-07-0582. *Clinical Radiol*, in press.

Emmerson, P.T. 1972. X-ray damage to DNA and loss of biological function: effect of sensitizing agents. p. 209–270. *In* Advances in Radiation Chemistry. Vol. 3.

Foster, J.L. & R.L. Willson. 1973. Radiosensitization of anoxic cells by metronidazole. *Brit J. Radiol* 46:234–235.

Foster, J.L. 1975. Mouse skin reactions after x-irradiation with the nitroimidazole radiosensitizer Ro-07-0582. *Brit J Cancer*, in press.

Fowler, J.F., J. Denekamp, A.L. Page, A.C. Begg, S.B. Field & K. Butler. 1972. Fractionation with x-rays and neutrons in mice: response of skin and C_3H mouse mammary tumors. *Brit J Radiol* 45:237–249.

Fowler, J.F., J. Denekamp, P.W. Sheldon, A.M. Smith, A.C. Begg, S.R. Harris & A.L. Page. 1974. Optimum fractionation in x-ray treatment of C_3H mouse mammary tumors. *Brit J Radiol* 47:781–789.

Fowler, J.F., P.W. Sheldon, J. Denekamp & S.B. Field. 1975. Optimum fractionation of the C_3H mouse mammary carcinoma using x-rays, the hypoxic-cell radiosensitizer Ro-07-0582, or fast neutrons. *Int J Radiat Oncol, Biol, and Phys*, in press.

Hall, E.J. & L. Roizin-Towle. 1975. Hypoxic sensitizers: Radiobiological studies at the cellular level. *Radiat Res*, in press.

Rauth, A.M. 1975. In vivo testing of hypoxic cell radiosensitizers. *In* Chemical Radiosensitization of Mammalian Cells. Proc. 5th Int. Congress of Radiation Research, 1974, Seattle. Academic Press, in press.

Scharer, K. 1972. Selective injury to Purkinje cells in the dog after oral administration of high doses of nitroimidazole derivatives. *Verk Deut Ges Pathol* 56:407–410.

Sheldon, P.W., J.L. Foster & J.F. Fowler. 1974. Radiosensitization of C_3H mouse mammary tumors by a 2-nitroimidazole drug. *Brit J Cancer* 30:560–565.

Sheldon, P.W. & A.M. Smith. 1975. Modest radiosensitization of solid tumors in C_3H mice by the hypoxic cell radiosensitizer NDPP. *Brit J Cancer* 31:81–88.

Sutherland, R.M. 1974. Selective chemotherapy of non-cycling cells in an in vitro tumor model. *Cancer Res* 34:3501–3503.

Urtasun, R.C., J.D. Chapman, P.Band, H. Rabin, C. Fryer & J.Sturmwind. 1975. Phase 1 study of high dose metronidazole on in vivo and in vitro specific radiosensitizing of hypoxic cells. *Radiology*, in press.

Withers, H.R. 1967. Recovery and repopulation in vivo by mouse skin epithelial cells during fractionated irradiation. *Radiat Res* 32:227–239.

A Biophysical Model
for Radiation Therapy

W. Pohlit, Ph.D.,
Max Planck Institute of Biophysics,
University of Frankfurt/Main,
Frankfurt, Germany

INTRODUCTION

In radiation therapy the destruction of a tumor embedded in healthy tissue is a process which involves the whole organism, and complicated reactions of the different organs occur. Some of the most important reactions, however, can be detected and analyzed by observing single cell reactions, with which this discussion is chiefly concerned. Most of the experiments discussed here have been conducted with a model system for eukaryotic cells, diploid yeast cells. The experiments with this model can be performed with high accuracy and with cells which are well synchronized in respect to the cell cycle. Furthermore, the biochemistry of the most important reactions of energy metabolism are well known for this system. As will be explained later, essentially the same results can be obtained in the same way with mammalian cells, and even mammalian tissue. Therefore, an extrapolation of these results to radiation therapy seems justifiable.

RADIATION EFFECTS IN G_1 CELLS

Yeast cells can be grown and the growth arrested in the G_1 phase of the cell cycle. They correspond to stationary mammalian cells, and although all experiments discussed here have been performed with such yeast cells, similar results can be obtained with mammalian cells.

G_1 cells are irradiated with increasing absorbed doses of sparsely ionizing radiation, such as high-energy x-rays or fast electrons. A dose-effect curve with a distinct shoulder is obtained when the vitality of the cells is tested immediately after irradiation by plating on a nutrient agar (Fig. 1). The irradiation can be interrupted after an absorbed dose, D_1, and can be continued after a certain time interval during which the cells stay in G_1. Then two different counter-reactions in the irradiated cells can be observed:

1. The number of vital cells increases if the cells are plated after a resting period in a buffer. Since no cell multiplication takes place during this period, a part of the lethal radiation damage must have been averted. This part of radiation damage is called "potentially lethal radiation damage" and this biological counter-reaction: "repair of potentially lethal damage."

2. During the same period of time the shoulder of the dose-effect curve, which was lost due to the irradiation, is now produced again. This counter-reaction should be called repair of sublethal damage, or "recovery from sublethal damage."

As will be discussed later in detail, for both biological counter-reactions energy in the cell is consumed. Therefore the presence of these reactions, or at least their time constants, strongly depends on the energy metabolism. The energy metabolism is quite different in normal healthy tissue and in neoplastic tumor tissue. The healthy tissue is well supplied with energetic molecules, such as glucose, and with oxygen for the efficient use of these molecules in ATP production. The tumor cells usually have a lesser supply of energetic molecules and oxygen, due to the abnormal growth of the tissue (Tannock, 1972).

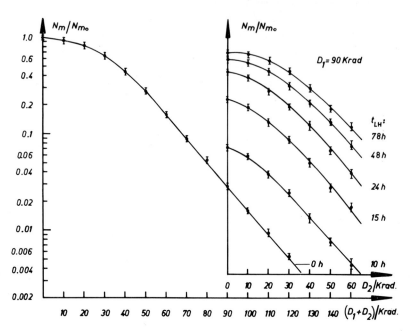

Figure 1. Dose effect curves for diploid yeast cells irradiated and kept in G_1 phase. The irradiation is interrupted at an absorbed dose of $D_1=90$ krad and is continued after a time interval t_{LH} during which the cells have been kept in buffer solution.

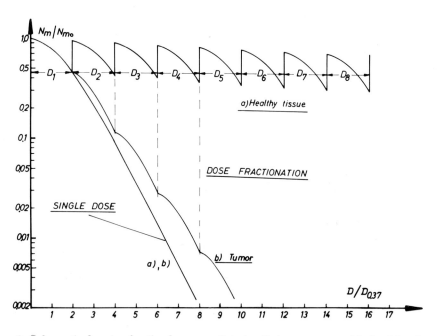

Figure 2. Schematic drawing for the decrease of vital cells in a tumor and in healthy tissue.

Therefore both counter-reactions can be assumed to occur much less frequently in the tumor than in healthy tissue. Under ideal conditions, in fractionated irradiation, the tumor cells will be destroyed following the single-dose curve (Fig. 2), but the healthy tissue will show repair and recovery during the radiation-free intervals. Even under ideal conditions the repair of potentially lethal damage will not lead to a survival of 1.0, since some irreparable damage must always occur, as will be explained later. In the patient under practical conditions, the tumor cells will also show some repair and recovery, thus reducing the difference between them and healthy tissue.

REPAIR OF POTENTIALLY LETHAL DAMAGE IN G_1 CELLS

The time course of repair processes in the cells can be followed experimentally. The cells are irradiated up to a certain absorbed dose, 120 krad, for example. Then the cells are plated after different intervals on a nutrient agar. As can be seen in Figure 3, the number of vital cells increases rapidly after a lag, but then levels off at less than 1.0.

Yeast cells (Strain 211p) which are genetically identical, but able to perform only fermentation and not respiration, are unable to repair potentially lethal damage. At first, it was thought that repair of potentially lethal damage might be linked biochemically with respiration. The following experiment clearly shows that this is not the case (see Fig. 4). When cells of the wild-type strain (211) which show repair of potentially lethal damage are treated with increasing concentrations of KCN, repair of potentially lethal damage is reduced and is completely inhibited at a concentration of 2 mmol/liter. But if concomitantly the energy metabolism via fermentation is increased by a supply of glucose, the repair function reappears.

In other experiments it has been shown that no respiration takes place under these conditions. The respiratory-deficient strain (211p) which normally in buffer is unable to repair potentially lethal damage, is able to do so if glucose is present in the buffer. Therefore, repair of potentially lethal damage is not directly dependent on respiration of the cell, but depends on the energy flux available for this counter-reaction. In the yeast cells used here, the energy is stored in an internal energy reservoir in the form of glycogen and trehalose. It is used either by fermentation or respiration for production of ATP, which is necessary for the maintenance of cell functions in the G_1 phase.

In the case of cells that ferment only, the energy flux from this reservoir normally is just high enough for maintenance, and no surplus energy is available for repair of radiation damage. The presence of glucose in the buffer, however, increases this energy flux sufficiently to allow repair processes. In the same way, repair capacity of tumor cells will depend on the energy supply and oxygen transport into the tumor. A reduced energy metabolism will reduce the repair capacity of tumor cells as compared to healthy tissue, and in this way increase the destruction of the tumor cells, as shown theoretically in Figure 2.

Such conditions, however, can not be assumed to be present in actual tumors. Even in hypoxic regions of the tumor, the energy supply via fermentation may be high enough to allow repair processes, as shown in Figure 3. It is therefore in our interest to reduce the energy production in hypoxic cells by inhibition of fermentation.

One possibility for achieving this is through the supply of desoxyglucose. This molecule is carried by active transport into the cell in the same way as glucose. During the course

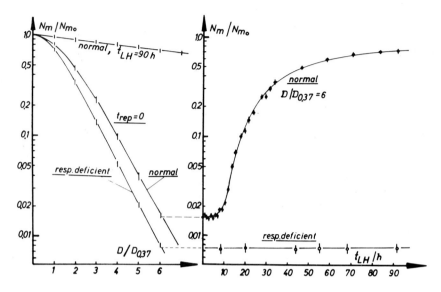

Figure 3. Dose effect curves for diploid yeast cells (normal wild type, 211, and respiratory-deficient strain, 211p) and time course of the repair of potentially lethal damage.

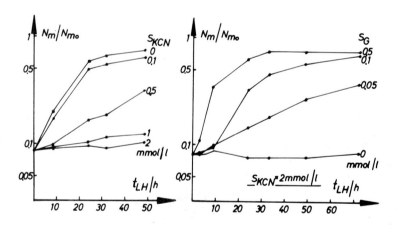

Figure 4. Time course of repair of potentially lethal damage in yeast cells (wild type 211) treated with KCN of various concentrations (left). Reappearance of repair function by increasing glucose concentration S_G (right).

of fermentation, in contrast to the normal glucose molecule, a product occurs which cannot be handled further by the cell enzymes. By feedback regulation, the accumulation of these molecules causes the fermentation in the cell to stop. This is a reversible reaction and can be bypassed by respiring cells. This reaction was checked first with the model system of yeast cells, as shown in Figure 5.

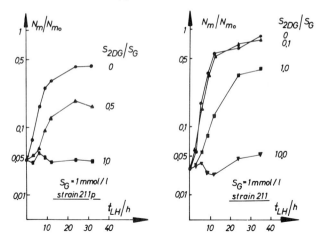

Figure 5. Influence of desoxyglucose on the repair ability in respiratory-deficient (left) and in normal (right) yeast cells.

In respiration-deficient cells (tumor cells with reduced oxygen supply) able to repair only in the presence of glucose, an increase in the concentration of desoxyglucose leads to a complete inhibition of repair at a relative level of $S_{desoxyglucose} : S_{glucose} = 1$. In respiring cells (healthy tissue with suitable oxygen supply) at this relative concentration of desoxyglucose, normal repair still occurs, since the remaining energy flux due to respiration is sufficient. Only at relative concentrations higher by a factor of about 10 is respiration inhibited. In this way, a differential radiation effect in healthy tissue and tumors *vis a vis* repair should be possible by introduction of such antimetabolites as desoxyglucose.

This has been tested in mammalian cells and in solid tumors in animals, using a method for quantitative measurement of dead cells in tissue which was developed by Porschen and Feinendegen (1969) and Pittner et al. (1973). In this method animals are fed with radioactive (^{131}I) iodinedeoxyuridine (IDU). This is incorporated, instead of thymidine, into the DNA of growing cells, so that radioactively labeled cells or radioactive tumors can be obtained. If a cell is killed, the IDU is released but not reused by other cells. The measurement of emitted gamma rays, therefore, is a measurement of vital cells in the organism of interest, and can be followed for the same tissue during a course of irradiation treatment. It could be shown that during fractionated irradiation, the number of vital cells in an irradiated tumor decreases, and that this decrease can be made greater by feeding the animal desoxyglucose. It should be emphasized that the reaction of desoxyglucose is completely reversible and that desoxyglucose is not toxic in the concentrations necessary to inhibit fermentation in the cell. Nonetheless, extensive experiments are needed to find the optimal concentrations and time schedule for application to patients.

With the same labelling method, reactions in different regions of tumors can be distinguished. If the animal is fed with ^{131}IDU during the initial growth of the tumor, the central part of the tumor emits gamma rays of ^{131}I (364 kV). The animal is later fed with ^{125}IDU and the outer parts of the growing tumor emit gamma rays from ^{125}I (35 kV), which can be measured separately. In this way the reduction of tumor cells in hypoxic and oxygenated regions of the tumor can be followed from outside the animal

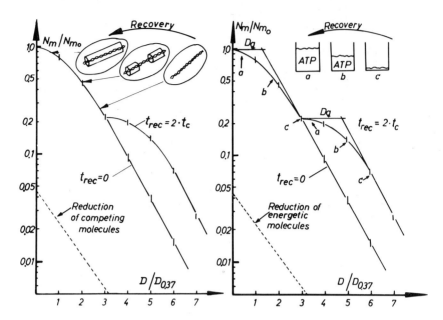

Figure 6. Two models for the explanation of the shoulder: Competing molecules (left) and fast repair (right).

by easy nuclear detection methods. Experiments of this kind are underway in our laboratory to test, during different irradiation regimens, the radiation sensitivity and repair capacity of these different tissues with their repair mechanisms inhibited via alterations in energy metabolism.

RECOVERY FROM SUBLETHAL DAMAGE

If living cells are irradiated with increasing absorbed doses, the slope of the dose-effect curve increases. In other words, the radiation sensitivity increases to a maximum at the end of the shoulder. The reappearance of the shoulder during a rest after irradiation means a reduction of radiation sensitivity. Until now there has been no clear molecular explanation for the reactions involved in the increase of radiation sensitivity during irradiation, nor for those involved in the counter-reaction after irradiation.

Two possible models have been developed in our laboratory and are under experimental testing now. The first model assumes that the changes (due to radiation) in the radiation sensitivity of an essential molecule in the cell (most probably the DNA) are brought about by a reduction of other molecules in the environment of this molecule, which compete in capturing radicals. With increasing absorbed dose the number of these competing molecules is decreased by radical reactions, and accordingly the probability of a reaction between a radical and the essential molecule increases. The cell recovers from this sublethal damage by reproduction of the molecules in the environment of the essential molecule (Fig. 6). Many experiments have been performed in our laboratory to test this hypothesis, but to date all of them have failed to support this model.

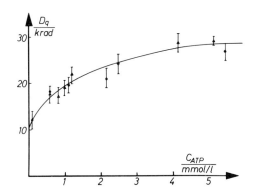

Figure 7. The length of the shoulder, D_q, in a dose effect curve for yeast cells as a function of ATP concentration in cells (▲ = wild type 211, ● = respiratory-deficient strain, 211p).

In the second model, it is assumed that the radiation reactions of the essential molecule occur with the same probability, independent of the total absorbed dose accumulated, but that a very fast repair of these reactions takes place as long as energy is immediately available. This energy, however, is increasingly consumed with increases in absorbed dose (Fig. 6), and at absorbed doses larger than the shoulder, this fast repair cannot take place. (Only the slow repair of potentially lethal damage, as described before, can take place with a time constant according to the energy production from the internal energy reservoir of the cell.) The length of the shoulder (D_q) should then be proportional to the energy immediately available for the cell.

In the same model system of yeast cells, the ATP pool has been measured as a function of absorbed dose. The ATP concentration in the cells indeed decreases with increasing absorbed dose, but not proportional to the absorbed dose. In further experiments, the content of the ATP pool of these cells has been altered by different biological and biochemical methods, and the correlation to the change in the length of the shoulder, D_q, is given in Figure 7. As can be seen, the ATP concentration is one of the limiting factors for D_q. At low ATP concentration in the pool, this energy determines the length of the shoulder. At a high ATP content another factor is limiting, the repair enzymes, for example. As can be seen from Figure 7, at extremely low ATP concentration, a minimum shoulder of D_q=10 krad exists, which seems to be independent of the ATP concentration and may be due to the ploidy of the cells. But a main part of the shoulder can be explained by a fast repair process, which is inhibited with increasing absorbed dose through the reduction of energy necessary for this fast repair.

Recovery from sublethal damage in molecular terms therefore means reproduction of the ATP in the pool. Experiments which show the filling of the ATP pool after irradiation also have been done in our laboratory (R. Reinhard, Doctoral Thesis, Frankfurt, Main, 1975). The dose-effect curve with a shoulder can be interpreted as a horizontal part, in which radiation damage occurs but is repaired immediately, and an exponential part, in which radiation effects cannot be repaired immediately. The curvature of the shoulder then must be attributed to decreased probability of repair with decreased ATP in the

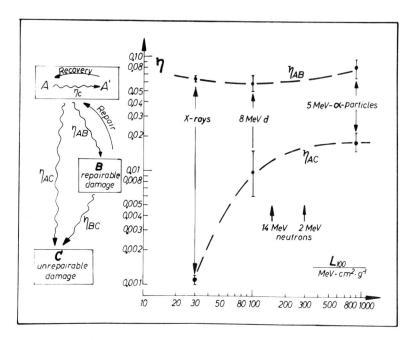

Figure 8. Schematic graph of the cybernetic model for radiation reactions in living cells, and dependence of the radiation reaction constants on linear energy transfer.

pool or to biological variance in the cell population with respect to the content of the ATP pool.

Since the content of the ATP pool and its replenishment after irradiation strongly depend on the energy metabolism of the cell, it can be influenced in a way similar to the slow energy production from the internal energy reservoir (glycogen) as described above. Indeed, the reappearance of the shoulder can be slowed down by supplying desoxyglucose to the cell.

In this way the energy metabolism of the cell influences both biological counter-reactions in a quite different way. The filling of the ATP pool is responsible for the radiation sensitivity, as expressed by the slope of the dose-effect curve, and the energy flux from the reservoir to the pool is responsible for the slow repair of potentially lethal damage. In this model a "sublethal" radiation damage does not exist per se; it is only an expression for the reduction of the ATP in the pool due to fast repair of potentially lethal damage.

CYBERNETIC MODEL FOR RADIATION REACTIONS IN LIVING CELLS

All the radiation reactions and biological counter-reactions described thus far can be used to construct a cybernetic model for radiation reactions in living cells (Fig. 8). Three different states of the cells are assumed: A, B and C. Cells that are able to proliferate when tested in a nutrient medium are in state A. The production of reparable damage transports the cell into state B with a reaction rate constant η_{AB}. Since in careful repair experiments some irreparable damage can always be observed, this is expressed by a state C into which cells are transformed from state A with a reaction rate constant η_{AC} and from state B with a reaction rate constant η_{BC}. By definition there is a biological counter-reaction from state B back to state A with a time constant ϵ_{BA}. Different radiation sensitivities of cells in state A can be expressed by a dose dependence of η_{AB}, which is influenced either by a reduction of competing molecules or by a reduction of the ATP pool. Both can be expressed by a radiation reaction rate constant η_c. The corresponding counter-reaction of recovery from this sublethal damage is included in this model as a time constant $\epsilon_{A'A}$. All these parameters of the cybernetic model can be determined experimentally, sometimes in many different and independent ways (Kappos and Pohlit, 1972). With use of this model, predictions about the dose-rate dependence and the survival of cells in different irradiation procedures can be made.

An application of this model to the description of radiation reactions produced with densely ionizing particles has shown that the reaction rate constant for irreparable reactions increases within a factor of about 20, if alpha particles are used instead of x-rays (Fig. 8), whereas η_{AB} stays nearly constant.

For radiation therapy this result means that the difference between the survival of healthy tissue and the survival of a tumor, as demonstrated in Figure 2, would be reduced after fractionated irradiation. Therefore, sparsely ionizing radiations, such as high-energy x-rays and fast electrons, seem to be of advantage in all cases where healthy tissue has to be irradiated during tumor therapy.

REFERENCES

Kappos, A. & W. Pohlit. 1972. A cybernetic model for radiation reactions in living cells. I. Sparsely ionizing radiations; stationary cells. *Int J Radiat Biol* 22:51–65.

Pittner, W., W. Porschen & L. Feinendegen. 1973. In-vivo-Messung unterschiedlicher Strahlenempfindlichkeit von Tumorzellen während des Zellzyklus; Zellmarkierung mit ^{125}I-Desoxyuridin. *Strahlentherapie* 145:161–168.

Porschen, W. & L. Feinendegen. 1969. In-vivo-Bestimmung der Zellverlustrate bei Experimentaltumoren mit markiertem Joddeoxyuridin. *Strahlentherapie* 137:718–723.

Tannock, I.F. 1972. Oxygen diffusion and the distribution of cellular radiosensitivity in tumors. *Brit J Radiol* 45:515–524.

Optimization of Energy and Equipment

Harold Elford Johns, Ph.D.,
Professor and Head,
Department of Medical Biophysics,
Ontario Cancer Institute,
Toronto, Ontario, Canada

In the last twenty years there has been a continual trend towards the use of high-energy x-ray machines in the 20–30 MV range for radiation therapy. We have already presented some evidence (Allt, 1969), based on Allt's report that the higher depth-dose of such machines is associated with an improved cure rate over that achieved with ^{60}Co in the treatment of carcinoma of the cervix. The major advantage of a linac over a betatron is the dose rate—1000 vs 50 rads per minute at a meter. Since the price of a linear accelerator increases with its maximum electron-beam energy, it is important to optimize the design of components to produce the highest effective x-ray energy at a given electron-beam energy and thus minimize the cost. In our earlier paper in this symposium, we showed that by changing the design of the target and the flattening filter, linacs could be made which produce the same depth-dose as the beam from a betatron. We trust that manufacturers will take these investigations into consideration in the design and production of future linacs.

There is one problem, however, which requires further discussion—the description of a linac by a number. For example, what does "Clinac 35" or "Clinac 18" mean? It is our understanding that the 35 and the 18 are merely the energy of the electron beam when the waveguide is unloaded, i.e., when there is negligible beam current. When the guide is loaded to produce a substantial output, the energy is less than this. Surely, the energy at which the machine is actually in use is the one which should be used to describe the machine. This problem should be resolved before any more confusion arises. Perhaps two numbers might be used—one to indicate the maximum useful electron energy, and the second to indicate the maximum photon energy. The notation might be linac (30:20). A better notation might be to specify for a standard field (10 cm X 10 cm at 100 cm) the properties of a beam in terms of three parameters: the surface dose, the depth of maximum dose and the percentage depth-dose at a depth of 15 cm. Thus our 22 MV betatron might be specified thus: (20%, 5.0 cm, 70%). This notation is unwieldly but would quite accurately describe the beam.

Now, we would like to address ourselves the difficult problem of determining the optimum energy of a linac. Is it 4, 6, 8, 10, 20, or 30 MV? To answer this question in an unbiased manner and give proof of the conclusion is by no means easy. As all of you know, the beam from a 4 MV linac is essentially the same as that from a cobalt unit. It is a matter of opinion as to which one we should use. However, it is not the purpose of the meeting to discuss the pros and cons of the two machines. Instead, we will leave the 4 MV linac out of our consideration and attempt to select the best energy in the range 6 to 30 MV. We will attempt to prove that the linac or betatron in the energy range 20 to 30 MV is the optimum machine for radiation. To do this, we will again refer to the report of Allt (1969) which, from a randomized study, compares the results with cobalt as opposed to a betatron in the treatment of advanced disease of the cervix.

Figure 1 shows Allt's survival curve for Stage III. We will now attempt to explain why Allt was able to report better results with the betatron than he did with cobalt. The arrangement of fields which were used in this treatment is shown in Figure 2. The cervix was treated with two opposing pairs of fields at an angle of about 55° from the vertical. It is seen that the high-dose region for the betatron is a rhomboid, and in this region the dose is more than twice as much as the dose at most points outside the field.

Figure 3 shows the distribution obtained with ^{60}Co, using the same four-field technique. The high-dose region is again rhomboid, but the region of high dose is smaller than

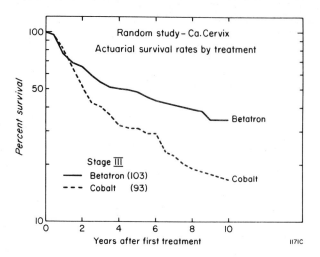

Figure 1. Comparison of survival data for Stage III cancer of the cervix treated with cobalt 60 and betatron radiation—(updated data from Allt's earlier study, 1969).

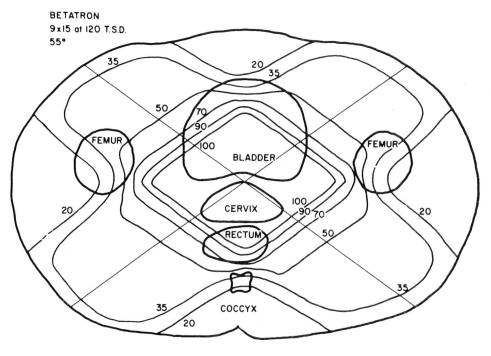

Figure 2. Isodose distribution from betatron operating at 22 MV. Field sizes 9 cm × 15 cm at SSD of 120 cm. Four-field technique for the treatment of Stage IIB and III cancer of cervix, according to Allt (1969).

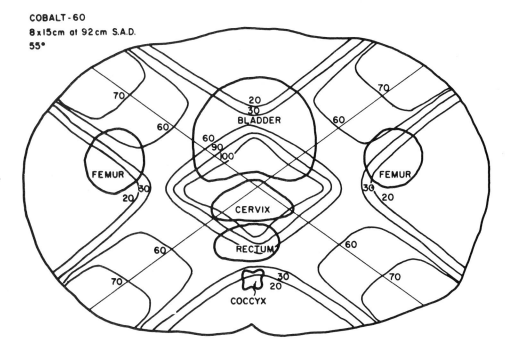

Figure 3. Isodose distribution for cobalt 60 technique used to treat Stage IIB and III cancer of cervix. Four fields, each 8 cm × 15 cm at the tumor at 92 cm source to axis distance, according to Allt (1969).

it was with the betatron, and the difference between the tumor dose and dose to normal tissue outside the tumor volume is not nearly as great.

If we look at the distribution of radiation in one plane, along a line at 55° to the vertical, we obtain the distribution shown schematically in Figure 4. The cobalt beam shows some skin sparing, a high dose at a point about 5 mm below the skin, a decreasing dose until one reaches the tumor volume, and then a rapid increase as one enters the region which is irradiated by the four beams. This is followed by an essentially uniform dose through the tumor. In contrast, the beam from the betatron has a smaller skin dose, shows less radiation impinging on normal tissues outside the tumor volume in region B. The field is slightly wider in the case of the betatron and delivers a slightly lower tumor dose. Originally, it was planned that these two beams should give the same dose in the tumor region, but in actual fact they differ by 6%. This was due to a mistake in calibration of the betatron. Remember, this study was started at a time when it was not clear how one should measure betatron radiation. The betatron field was about 1 cm wider, and this was realized after the study was underway. It was decided to leave the fields as they were since the betatron patients were tolerating the wider field with no complications, whereas the cobalt patients were already in some trouble with respect to complications, and it would have been unwise to increase the field size to match that of the betatron.

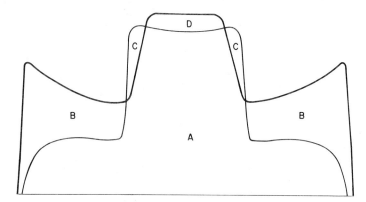

Figure 4. Schematic diagram showing the dose in a plane perpendicular to the diagrams of Figs. 2 and 3.

There were several other factors which were not identical in the two treatments. With the betatron, only one field was treated per day, whereas with the cobalt unit, four fields were treated per day, so that there was a difference in fractionation, the effect of which cannot be predicted with our present knowledge of radiobiology. There was a difference in the way the patients were positioned in the beam. For the cobalt treatments, the patient was supine for all four fields, and the machine was rotated around the patient to deliver the four beams of radiation, whereas in the betatron, the patient was supine for half the treatments and prone for the rest of them. This could introduce differences between the two modes of treatment, since internal organs of the patient could shift in different ways in the two treatments.

However, we believe the main difference between the techniques is explainable using Figure 4. The healthy tissues in region B are protected using the betatron technique. If we interpret Allt's data in this way, then the question immediately arises, what would have happened had we used a 10 MV linac beam, instead of a 25 MV betatron. At what energy does one have an increase in cure rate—at 10, 15, 20 MV, or what energy? This, of course, is something one cannot answer unequivocally. To look at this problem further, let us consider the distribution of radiation produced by two opposing pairs of beams as a function of energy. In Figure 5 we show the net result of placing two beams in opposition 25 cm apart with the dose at depth 12.5 cm normalized to 100. The top curve is for cobalt 60, and shows that when the beams are normalized to give the same dose at the tumor, the tissue just below the skin surface receives 126% of the radiation received in the tumor region. As the energy is raised to 10 MV, the dose contour becomes almost flat, and finally at 32 MV it shows considerable saving of the tissues in front and behind the tumor volume.

Now, one might say that one should not use opposing beams in high-energy radiation therapy. However, regardless of how one arranges the beam, for the same tumor dose, the dose to normal tissue just below the skin for cobalt is higher than for 32 MV radiation as shown in the bottom part of this diagram, where single fields have been considered. Beyond the tumor, the cobalt beam is better than the linac, because the dose falls slightly faster. So one has the problem of deciding whether you gain more in front of the tumor than you lose in the region beyond the tumor as the photon energy is increased.

Optimization of Energy and Equipment

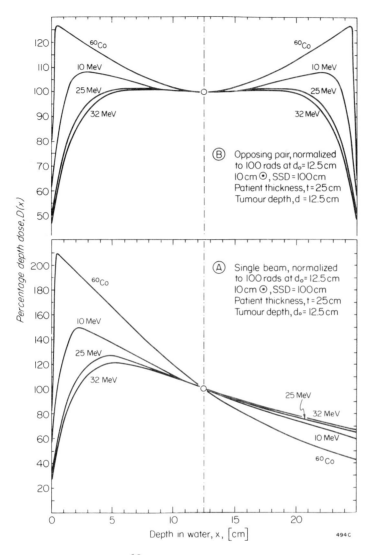

Figure 5. Depth-dose curves for ^{60}Co, 10, 25, and 32 MV beams normalized to 100 rads at a depth of 12.5 cm. (A) single beam; (B) opposing pair. Field size: 10 cm diameter at an SSD of 100 cm. The data are plotted for our experimentally determined depth-dose using the optimum target and flattening filter combination at a given energy. (Reprinted from Podgorsak et al., 1975.)

One way to characterize this problem mathematically is to calculate the integral dose for a given tumor dose. We realize that most people believe that integral dose bears very little relation to any practical problems in radiation therapy. However, it is a mathematical way of expressing the amount of energy that is absorbed by the patient. The ratio of the integral dose to the tumor dose is a good parameter to compare two types of radiation, and this ratio should be minimized.

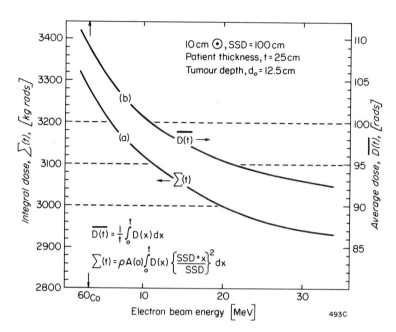

Figure 6. Integral dose: (a) average dose along the central axis (b) as a function of the electron beam energy, when a point at the center of a 25-cm thick patient is given a dose of 100 rads. Field size: 10 cm diameter at an SSD of 100 cm (Reprinted from Podgorsak et al., 1975).

The integral dose received by a patient of thickness 25 cm, when a point at depth 12.5 cm is given a dose of 100 rads using a 10-cm circular field, is shown in Figure 6. It is apparent that increasing the photon-beam energy reduces the integral dose per unit tumor dose. This is another way of saying that as the photon energy is increased, more radiation is delivered to the tumor and less radiation to the normal tissues around it. Calculations depend, of course, on what size tumor is assumed and the thickness of the patient. From Figure 6 it is evident that for energies above approximately 25 MV the curve becomes essentially flat, so very little is gained in going to energies above 25 MV. This is also seen in Figure 5, where the difference between 25 and 32 MV is very small.

Now what is the optimum energy? We believe it is about 25 MV. Higher energies produce little improvement in depth-dose, while lower energies increase the dose to normal tissue relative to the tumor. In the case of betatrons, there may be an advantage in using higher energies. The higher energy only improves the depth-dose marginally, but can give a large increase in yield.

CONTROL OF PHOTON BEAM

We will now discuss the problem of optimization of equipment. One of the problems all of us worry about in the use of high-energy machines is to ascertain that the beam is, in fact, delivered in the way in which the manufacturer and the radiation therapist intended.

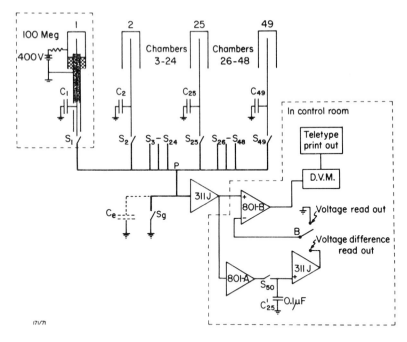

Figure 7. Schematic diagram of the circuit for the matrix dosemeter. S_1 to S_{49}, S_g and S_{50} are reed relay switches. C_1 to C_{49} are capacitors to store the charge liberated in ion chambers 1 to 49. Measurements are made relative to chamber number 25, which is placed at the center of the field. (Reprinted from Johns et al., 1974.)

One uncertainty users of linacs and betatrons have in common is that of whether the dose is uniform over the treatment field. The flatness of high-energy beams depends very critically on the positioning of the flattening filter in the beam. The higher the energy, the more difficult this problem becomes, and the more care one must take in making sure that the beam is lined up correctly. We have found that a good way to measure the flatness of the beam is to use our matrix dosemeter (Johns et al., 1974). This is a device which measures the dose at 49 points in the radiation field. These points can be chosen at random, but we have chosen to put them on a rectangular grid, with seven detectors along each side. The principle of the matrix dosemeter is shown in Figure 7, where we show four of the 49 ion chambers connected to an electronic circuit.

The 49 chambers are mounted on a jig at the end of the linac, and each of them is discharged by momentarily closing the grounding switches, S_1, S_2, etc., and the switch S_g. All chambers are then given an exposure, and then the charge stored on each of the capacitors C_1 to C_{49} is measured by a solid-state amplifier circuit, and is either presented on a teletype or a teletype printout, which may then be processed with a computer to give a distribution of radiation. We built this device originally to check the Varian linac, but found it so useful that we now have one for each of our two high-energy machines. These are used routinely each morning before the machines are used in the treatment. Typical results are shown in Figures 8 and 9.

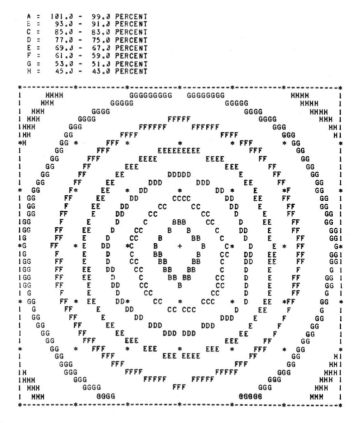

Figure 8. Isodose display from matrix dosemeter for "Clinac 35" with flattening filter removed. (Reprinted from Johns et al., 1974.)

Figure 8 shows the isodose display for a beam in which no flattening filter is used. In lining up the linear accelerator, it is important to get a distribution of this kind without the filter in place to assure that the electrons are coming down the axis of the waveguide.

If one now puts the filter in place, and gets it adjusted properly, one obtains a uniform distribution as shown in Figure 9. The distribution shown here is really quite good, the difference between A and B being only 4%. We have used this matrix dosemeter continuously since we have put our linac in operation, and we find that it is a very sensitive diagnostic indicator of the behavior of the machine. When something goes wrong, the first thing one does is put the matrix dosemeter on the machine and take one reading.

The state of the art in solid-state amplifiers, of course, changes very rapidly, and it is now possible, we believe, to create a matrix dosemeter which looks at the beam continuously and would enable one to get a profile at any time during the treatment. This would be a big advantage; ways of doing this are now being investigated.

OPTIMUM HEAD DESIGN

Finally, we will discuss possible designs for the head for linear accelerators, which we think would yield an optimum dose distribution. We consider, first, the high-photon beam. In most linacs the electron beam from the waveguide impinges on the target essentially only in one direction, that is, along the axis of the machine. Of course, there is a slight angular spread of the electron beam at the target, but it is quite small compared with the spread required to give a 35 cm X 35 cm field at 1 meter. It would be advantageous if the electron beam impinged on the target along the sides of a cone, with the angle of the cone slightly less than the angle required to reach a field of 35 cm X 35 cm at a meter. To our knowledge, this has never been done, but we feel sure it could be done. If a conical beam could be obtained at the target, the flattening filter could be made very much thinner, because now photons would be directed towards the edge of the field as well as down the axis. A flattening filter would still be required.

If the target is thick, it should be made of aluminum, or some other material with a low atomic number. Instead of aluminum, a more dense material such as copper would probably serve almost as well. The best design would be to have a thin target and remove the electrons with some kind of magnetic field. To our knowledge this has never been done. However, it is not clear that this would improve the beam very much, as the work

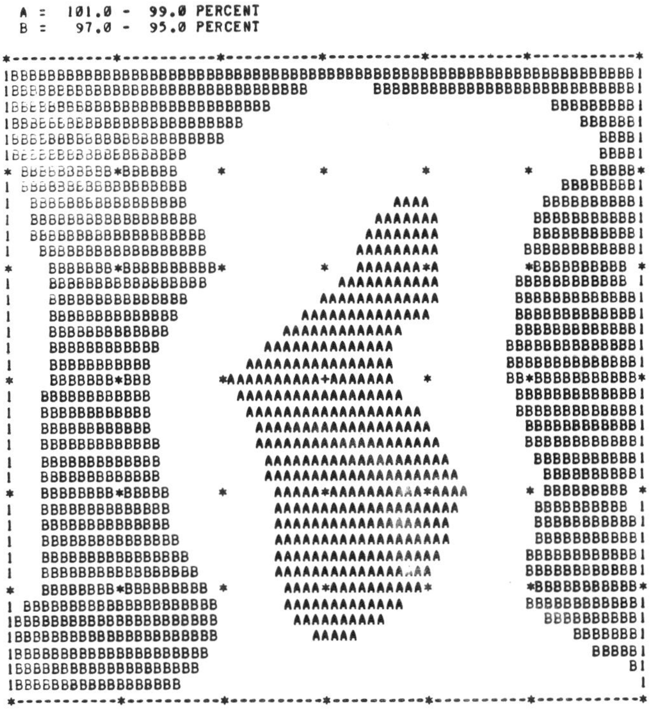

Figure 9. Flattened linac beam. The numbers in the table give the actual readings of the 49 chambers. A represents 101.0 to 99.0%, B–97 to 95.0%. (Reprinted from Johns et al., 1974.)

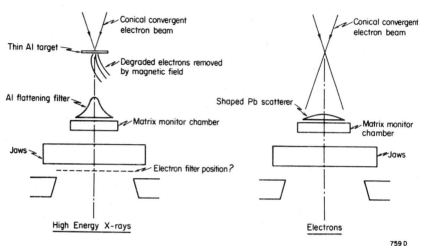

Figure 10. Schematic design of head for linear accelerator used in the photon mode (left) and electron mode (right).

of the group in Ottawa (Sherman et al., 1974) has shown that a thin aluminum target and a thick aluminum target give essentially the same spectral distributions of radiation. We place a matrix monitor chamber below the flattening filter, arranged so as to continuously display the distribution at the control panel. The beam-defining equipment is well designed and we do not have any suggestions for improving it.

Finally, there should be an electron filter to minimize contamination of the beam. The optimum design of this filter is not known, although our experiments indicate that lead is slightly better than any other material for the filter. We have found also that it should be placed near the flattening filter. One position might be between the jaws of the collimator. Since one would like to have a visual indication of the field size, it is probable that the electron filter would have to be introduced automatically when the beam is turned on, but is out of position during the time the patient is being set up for electron therapy.

The redesign of the head of a linear accelerator is by no means a trivial problem. It is hoped that in the future the target and flattening filter will be made of the proper material so that the user does not have to spend a lot of money and time reconverting the head to yield the beam characteristics required.

On the right-hand side of Figure 10, we show possible designs for a head for electron therapy. Again, a conical-type, converging, electron beam would be advantageous, and this would automatically spread electrons to the edges of the field. A shaped lead scatterer would also be required, and it would be placed just above the matrix monitor chamber. With these designs, the full potential of linear accelerators in the treatment of deep-seated tumors should be achieved.

ACKNOWLEDGMENTS

The authors take pleasure in acknowledging the financial assistance of the National Cancer Institute of Canada and the Medical Research Council of Canada. We are indebted to Dr. Ervin Podgorsak and Mr. Duncan Galbraith for their help in taking many careful measurements, and to Dr. W.E.C. Allt for making his updated, unpublished, survival statistics available.

REFERENCES

Allt, W.E.C. 1969. Supervoltage radiation treatment in advanced cancer of the uterine cervix—A preliminary report. *Canad Med Assoc J* 100:792–797.

Johns, H.E., J.A. Rawlinson and W.B. Taylor. 1974. Matrix dosemeter to study the uniformity of high energy x-ray beams. *Amer J Roentgenol* 120:193–201.

Podgorsak, E.B., J.A. Rawlinson and H.E. Johns. 1975. X-ray depth doses for linear accelerators in the energy range from 10 to 32 MeV. *Amer J Roentgenol* 123:182–191.

Sherman, N.K., K.H. Lokan, R.M. Hutcheon, L.W. Funk, W.R. Brown and P. Brown. 1974. Bremsstrahlung radiators and beam filters for 25-MeV cancer therapy. *Med Phys* 1:185–192.

Future Uses for High-Energy, Low-Let Radiation

William E. Powers, M.D.,
Professor and Director,
Division of Radiation Oncology,
Mallinckrodt Institute of Radiology,
Washington University School of Medicine,
St. Louis, Missouri

Radiation therapy cures cancer, yet all agree that improvements in radiation therapy, using any of the available mechanisms of delivering radiation, can be made with appropriate biology, physics, and clinical understandings and applications. While the alternative methods of treatment with high linear energy transfer (LET) radiation are being developed, it is quite appropriate that we formalize and reinterpret the expected results and the mechanistic processes of radiation therapy success and failure with the modalities, equipment and personnel we have available today.

Those of us who have megavoltage units with energies over 10 MeV are convinced that high-energy photon (25 MeV) and electron (6-21 MeV) beams are more effective than lower-energy beams for both cure and palliation. We have not convinced all of our colleagues in radiation therapy and in other branches of the practice of medicine of our convictions.

TUMOR LOCALIZATION

No matter what energy or modality of radiation is utilized for radiation therapy, we must develop skill in determining the anatomic extent of the tumor cells. Conventionally we determine tumor extent, and thus the volume to be treated, by a combination of history and physical examination including x-ray procedures and laboratory studies. We add to these clinical findings in the individual patient a prediction of the subclinical extent of the tumor based on the accumulated experience concerning patterns of spread with tumors of similar histology and location. The newer methods—ultrasound studies and isotope procedures, as well as the use of specific tumor markers such as the immunoreactive test—will allow a better definition of tumor extent and improved treatment planning. Computer-assisted tomography of whole body sections provides remarkable detail and definition of tumor extent and will be mandatory for treatment planning in many body regions. In addition to determining the extent of the grossly apparent tumor, it will become necessary for us in our practice to estimate the anatomical distribution and density of tumor cells with the construction of the "isotumor cell density contours."

Our practice of delivering varied doses of radiation to portions of patients represents an implicit presumption of a variation of tumor cell density throughout the region treated. This is implicit in our practice, but is not defined or specified in our patient records. If ever available, a method of estimating the tumor cell density through the various tissues to be irradiated would be of considerable value; for practical purposes, we make estimates. For example, within the primary tumor of an epidermoid carcinoma of the cervix, there are probably 10^9 tumor cells per cubic centimeter of the primary tumor. This tumor tissue is mixed with a lesser component of normal tissue and stromal structures. In the immediate parametrial area where there is questionable tumor extension, the tumor-cell number may be in the range of 10^7-10^8 tumor cells per cubic centimeter. In the peripheral portion of the pelvis or in the pelvic lymph nodes, the tumor cell burden may be 10^9 cells per gram in lymph nodes that are definitely involved, as determined by lymphangiographic or pelvic findings. Alternatively, in those nodes that appear to be negative on the lymphangiogram, and in the absence of palpable pelvic extension, there may nevertheless be as many as 10^6 or 10^7 tumor cells per cubic centimeter. Our administration of high-dose intracavitary therapy into the cervical primary tumor with the acceptance of a lesser dose in parametrial tissues has "operationally" accepted this inhomogeneity and variation in tumor cell distribution. It is probably time that we begin a formalized declaration of the tumor-cell density in order to logically develop a predicted requirement of radiation dose for curative treatment.

DOSE VERSUS CONTROL

Observations of the steepness of the curve relating the "percentage of local control" to the "dose" of radiation have been made and we accept the proposition that an extremely steep dose-response curve exists (Herring, 1975). The frequency of necrosis has also been related to the dose delivered, but in only a few cases and with less firmly established data. These observations lead to the clear understanding that there may be a small difference between the dose required for the cure of a high proportion of patients and the dose that will result in complications. These estimates lead to the conclusion that about 10% inaccuracy in dose delivered may result in either failure or complications.

These observations lead logically to the presumption that careful treatment planning will be required to fit the dose to be delivered to the predicted tumor volume and that both must be carefully determined so as to preclude: (a) underdosage in a portion of the predicted tumor or (b) overdosage to the adjacent or interstitial normal tissues.

Further, for the appropriate dose to be delivered to the irregular tumor in an irregular patient, the best treatment will probably require a variety of beams (possibly both photons and electrons), a variety of energies, careful beam direction and also a considerable capability and sophistication in beam shaping and modification so that the high dose of radiation conforms to the high density of tumor cells and yet an excessive dose is not delivered to sensitive structures.

In addition to the treatment planning and the selection and utilization of a variety of energies and types of beams with suitable field-shaping and beam-attenuation devices, it will be imperative that there is precise beam localization and delivery of the dose of radiation over the prolonged fractionated course of therapy. This will require precision in original set-up of the fields, later re-positioning of the patient for the many sessions, and immobilization during each exposure period.

CONFORMATION THERAPY

The distribution of dose so that treatment will be effective with a high probability of local control and a low probability of necrosis calls for a careful fitting of the dose gradient to the gradient of tumor-cell density (Takahashi, 1965). This adjustment of dose to the tumor requires: (a) a formalized definition of the extent of the tumor, (b) an optimized treatment plan, and (c) the equipment available for delivery of the optimized dose. Among the problems of dose optimization is that we have not established the characteristics of the "best treatment plan." It is difficult, if not impossible, to seek out the best treatment plan with either the computer or our past experience. However, we can approximate the "best treatment plan." We profit considerably from the experience of the excellent observer and the creative originator of treatment plans that have high degrees of success and low incidence of failure.

In the past our specifications of dose have been inadequate, as they have generally included only the given dose and the maximum tumor dose. A mechanism for better specification of dose and better comparison of treatment plans will be the profile of dose across a selected, important diameter of the patient. For example, the dose profile across a patient 14 cm thick (Perez et al., 1976), indicates relatively small variation in mid-diameter dose when the patient receives parallel, opposing, 4- or 35-MeV beams; however, there is a considerable difference in the dose at the surface of the patient. In Figure 8 in the same chapter, there is a considerable difference between the results with 10- or 25-MeV linac beams and those with cobalt-60 or 4-MeV linac beams in delivering a mid-diameter dose in a patient of 25 cm thickness.

Figure 13b, in the same chapter, indicates dose profiles along an anterior-posterior diameter through the mid-portion of the pelvis of a patient with carcinoma of the prostate and demonstrates that the weighted parallel, opposed portals on the betatron produce a more appropriate dose distribution with a lower dose to tissues anterior and posterior to the prostatic tumor and its pelvic extensions than does the "Clinac 35" also operated at 25 MeV. In the comparison of these dose profiles, it is possible to subtract the difference in dose visually and determine if the excessive dose or underdose is in a critical area of the patient. Present-day computer systems permit a series of dose profiles to be constructed at the will of the physician. He may use this "dose profile" technique to compare and contrast the different treatment plans and thus approach the optimum plan. This "dose profile" also identifies the maximum dose within the patient along the diameter selected.

In the attempt to optimize dose using dose profiles, we have found that in almost every type of patient a combination of high-energy photons, either alone or with electron beams, has permitted a more appropriate dose distribution for the tumor under consideration than cobalt-60 or 4-MeV beams used alone. Shaping of the perimeter of these beams is not difficult, and the beam margins are consequently quite sharp. Beam direction with the older units, such as the Allis-Chalmers betatron, is considerably more difficult than with the more flexible new betatrons or accelerators. The flexibility of these units, allowing selected positioning, beam direction, and beam modification, is now added to the benefit of the variation of dose profiles available with the various available beams.

DESIGN OF NEW BEAMS

We must be quite careful in the design of new equipment and the persuasion of equipment companies to manufacture new forms of equipment to correctly and appropriately specify our desires and demands. In the selection of an "ideal megavoltage therapy unit," the primary considerations were: (a) a very high energy beam; (b) good field flatness; (c) large fields, and (d) a high dose rate. We have obtained these results in the various 35-MeV linear accelerators ("Sagittaire" and "Clinac 35"). This attainment, however, has been at some significant cost in the depth of the maximum dose and in the depth of the 50% dose, as compared to the thin-target betatrons which give a deeper D_{max} and a deeper D_{50} than do the accelerators. We must, therefore, remember to specify the characteristics of the beam that we desire for *radiation therapy practice*. When we seek help from physicists, engineers, and manufacturers to translate our *clinical* needs into *engineering* and *physics* terms, we must be careful that we continually make sure that all disciplines understand the implications of our message. That we have not accomplished this in the past is attested by the presentations here of data about accelerators that meet our specifications for field size, dose rate, and field flatness, but fail to meet our expectations for depth of D_{max} and D_{50} (Almond et al., 1970; Rawlinson and Johns, 1973; de Almeida and Almond, 1974; Perez et al., 1976; Purdy et al., 1975; Velkley and Purdy, 1975).

FAILURE VERSUS CONTROL

We have considered that the primary reason for failure of radiation therapy has been the presence of an anoxic or hypoxic (and therefore resistant) subpopulation of tumor cells that have the capacity, after the completion of the radiation therapy, to repopulate the tumor.

Alternative causes of failure should be sought. An evaluation of a population of patients who received preoperative radiation therapy and had resection performed approximately 2 weeks after the completion of delivery of 3,000-rads tumor dose indicated that in a significant population of these patients total disappearance of their tumors occurred, as estimated by detailed microscopic examination of the pathological specimens (McGavran et al., 1964). We would predict that none of these patients would have been cured by the 3,000 rads and that split-course therapy (the original 3,000 rads represents the first half of a split-course) would also fail to cure a large portion of these patients. Since good healing of epithelium occurred and the mass was almost always markedly decreased, we should expect "reoxygenation" of the small number of remnant cells (cells not recognized by the pathologist, but surely present). Since split-course therapy doesn't always work, we should continue to search for other mechanisms of tumor recurrence in addition to *hypoxic cells*.

We should look at other mechanisms of tumor control. We have presumed that every tumor cell is able to repopulate a tumor; however, there may be a small number of *tumor stem cells* within the tumor mass. This would allow moderate doses of radiation to be curative, as in the case of Hodgkin's disease or seminoma, where moderate doses of radiation cure very large masses of tumor. These observations should lead us to investigate the possibility of (a) *sensitive tumor cells* or (b) a low proportion of tumor cells being actually "tumor stem cells." We should not let our preoccupation with "hypoxic tumor cells" fix us too firmly into "only" high-LET studies and thus preclude our examining other alternatives and possibly profitable areas of investigation.

HEALING AND NECROSIS

While high-energy, low-LET radiations allow us to control tumors better, we still produce damage to a variety of normal tissues. We presume healing takes place in irradiated tissue from either remnant surviving cells in the radiation field or migration of normal cells into the field. I continue to be amazed at the capability of the body for healing and also at my own incapacity to document and determine the magnitude of contribution of these two alternative mechanisms of healing. If we understood the mechanism better, we might be able to improve healing processes. We presume that necrosis is due to either severe depopulation of the normal stem cells or late effects of small-vessel damage. Again, we do not know the magnitude of contribution of these two alternative mechanisms.

In observing the specimens from patients receiving the equivalent of the first half of the split-course radiation, we note a significant amount of interstitial fibrosis in the tissue replacing the tumor. This fibrosis is somewhat loose and areolar at the time of the operative procedure. There is almost always good mucosal healing in the areas not totally denuded by the presence of tumor. A significant, immediate, normal-tissue response has taken place in the early interval after the moderate doses of irradiation and in a short time-span. A significant portion of the immediate post-irradiation early effect is this extensive fibrotic replacement of tumor and normal tissue destroyed by the radiation therapy. It may be that late condensation of this fibrosis into dense avascular connective tissue is the primary cause of the late damage rather than the small-vessel damage. Our preoccupation with small-vessel effects may lead us into incorrect mechanistic evaluations of the healing processes and away from potentially important alternative processes of healing and of late necrosis.

In addition to immediate or late effects that result in obvious scarring and necrosis, we must begin evaluating subclinical damage to normal tissues. In the vagaries of life, each of our patients will likely suffer a variety of disease processes involving critical organs. The possibility certainly exists that radiation further damages these diseased tissues. Clear-cut cases occur in the onset of osteoradionecrosis of the mandible after an extraction, or skin breakdown after minor trauma. Less clear, but potentially more significant, are the cases of patients whose cardiac or renal disease (statistically frequent diseases in unirradiated patients) occur in heart or kidney that was incidentally included in a treatment field. It is extremely difficult to measure subclinical disease. We have considerable problem predicting the degree of "illth" developing as a result of irradiation ("illth" is defined as the opposite of health). We do not have adequate techniques to measure the degree of damage to normal tissue when it is less than the damage that results in necrosis or near-total dysfunction. And yet, we must compare and contrast various forms of radiation therapy on the basis of the "illth" that they produce in the local tissues.

SCORE FUNCTIONS (BENEFIT-DEFICIT; RISK-REWARD)

All the practice of radiation therapy is based on the observations that we can cure patients with preservation of function and structure of most of the organs in the beam path. For the future, we need better score functions of the extent and frequency of necrosis and/or subclinical disease. We also have to put a value or *score* on: (*a*) cure and (*b*) preservation of organ and function.

In addition to these direct and immediately apparent damaging effects of radiation therapy, other effects occur. Quite appropriately we face several new and important challenges as to the frequency and significance of several less clear-cut and subtle ill effects. These include the assay of the degree and significance of *immunosuppression, carcinogenesis, genetic perturbation,* and *growth retardation*. All four of these events will occur to a significantly increased extent as we improve our cure rates and prolong the lives of our patients.

Our score must include not only relatively late effects at 1 and 2 years, when most necrosis takes place, but also much later effects occurring at 10, 15 and 20 years after successful therapy.

HIGH-ENERGY VERSUS MODERATE-ENERGY RADIATION

We have not demonstrated to the satisfaction of the administrators, physicists or practitioners who have cobalt units and 4-MeV accelerators the significance and effectiveness of the betatrons and high-energy accelerators. While we may, in an individual patient or a small group of patients, be able to demonstrate a better treatment or a better result, these may be attributed to superior skill or superior selection of material.

The study from Princess Margaret Hospital showing increased survival and decreased complications in patients treated on the betatron is outstanding. Such studies and their results are needed.

A general demonstration of the increased effectiveness of high-energy therapy units must be made if we are to justify purchase and installation of this very expensive equipment and to justify to the equipment manufacturers and insurance carriers the marketability of the product.

HIGH-ENERGY, LOW-LET VERSUS HIGH-LET RADIATIONS

Predictions presume a better knowledge of the future than is given to most of us. However, we have heard presented at this meeting the series of alternative improvements that can be accomplished with high-energy, low-LET radiation even in a "high-LET era." These improvements, and the biology and physics implied, need study, and fortunately with the emergence of the "high-LET era," the biological and clinical studies of low-LET, high-energy radiation will be accomplished in a more formalized and complete manner than has been possible in the past. Both the high- and low-LET programs and patients will benefit.

Alternatives to high-LET radiation include modifications of time-dose fractionation. We have habitually followed certain time-honored and probably quite effective time-dose-fraction schemes to the exclusion of investigations of alternatives. Variations based on animal studies, computer simulation models, and accumulating clinical experience are being developed and these projects may allow alternative fractionation systems designed to fit a specific type of tumor in a specific region of the body. Dose-shaping to fit the tumor will be required in whatever form of therapy we utilize—whether "high-LET" or "high-energy low-LET." However, dose shaping will be more difficult for neutron-beam therapy and not simple for pion therapy. The use of radiation sensitizers for hypoxic cells or resistant-population cells may add a significant component to our accomplishments. An increased interest and effort in the use of interstitial and intracavitary therapy sources provides a means of applying high doses of radiation to the tumor with sharp gradients of dose fall-off in the adjacent normal tissues. These techniques deserve wider usage, as this is an appropriate extension of the concept of fitting the high dose to the irregular volume of high density of tumor cells under direct vision. The use of shorter moderately-lived isotopes as permanent implants with good dose distribution calculation and good patterning of the source placement is an exciting step towards improved care. External beam-shaping to fit the interstitial and intracavitary doses will allow a graded dose to be delivered with a low probability of injury. The use of hyperthermia to modify the local effect of radiation in a selected volume of the patient may be contributory if good control of the volume of increased temperature can be developed.

Combined modality therapy, making use of surgical resection to remove the primary tumor (high tumor-cell density), radiation therapy to destroy regional or local extensions of the tumor (preoperative or postoperative), and chemotherapy to kill disseminated cells, has been effective in a number of tumors in children. Variations of these techniques are being employed with variable success in tumors of adults. Study of both the successes and failures of the concept of combined modality therapy must be undertaken. We also need to learn better how to use combined modality therapy in concurrent, as well as sequential, applications of the agents.

SUMMARY

We face an exciting future with alternative solutions available in clinical studies, biological investigations, and physics problems to be met. Most of these alternatives provide the hope of optimistic and positive solutions with increased effectiveness and applicability of high-energy, low-LET radiation for improved cancer palliation and improved rates of cancer cure.

REFERENCES

Almond, P., E. van Roosenbeek, R. Browne, J. Milcamp & C.B. Williams. 1970. Variation in the position of the central axis maximum build-up point with field size for high-energy photon beams. *Brit J Radiol* 43:911.

de Almeida, C.E. & P.R. Almond. 1974. Comparison of electron beams from the Siemens betatron and the Sagittaire linear accelerator. *Radiology* 111:439–445.

Herring, D.F. 1975. The consequences of dose response curves for tumor control and normal tissue injury on the precision necessary in patient management. *Laryngoscope* 85:1112–1118.

McGavran, M.H., J.H. Ogura & W.E. Powers. 1964. Small-dose preoperative radiation therapy. A preliminary report based on some histological observations of thirty resected epidermoid carcinomas of the upper respiratory and digestive tracts. *Radiology* 83:509–519.

Perez, C.A., J.A. Purdy, A. Korba & W.E. Powers. 1976. High-energy X-ray beams in the management of head and neck and pelvic cancer. p. 217–241. *In* Kramer, S., Suntharalingam, N. and Zinninger, G. (eds.) High-Energy Photons and Electrons: Clinical Applications in Cancer Management. John Wiley and Sons, New York.

Purdy, J.A., F.R. Zivnuska, D.E. Velkley & D. Keys. 1975. Percent depth dose studies of the Clinac 35 linear accelerator. *Med Phys* 2:165 (Abst.).

Rawlinson, J.A. & H.E. Johns. 1973. Percentage depth dose for high energy X-ray beams in radiotherapy. *Amer J Roentgenol* 118:919–922.

Takahashi, S. 1965. Conformation radiotherapy: rotation techniques as applied to radiography and radiotherapy of cancer. *Acta Radiol* (Suppl.) 242:142.

Velkley, D.E. & J.A. Purdy. 1975. Depth of maximum dose for megavoltage photon beams. *Med Phys* 2:166 (Abst.).

Discussion

Dr. Powers: First, I would like to ask the panelists to ask each other questions. I particularly want the biologists to intepret the biomathematical model. Does Dr. Johns agree that a dose profile is a fairly good representation of what we're doing?

Dr. Johns: I gather you want to specify the dose along a line.

Dr. Powers: One or a number of lines; we should be able to dissect them.

Dr. Johns: It's an excellent idea. We are going to have scanners very shortly which will help in the dosimetry problem. The field is moving very, very fast, and scanners will be used in both treatment planning and diagnosis.

Dr. Powers: In defining the tumor extent with the computational systems presently available, we can already do a fair degree of treatment planning; with the new units we can probably plan a fairly good delivery of the dose to the tissues. We sometimes forget that it is not at all adequate to simply prescribe a dose. In radiation therapy, with our schemes of fractionation, we must presume significant precision in administering the daily doses; we are totally dependent on a variety of skilled individuals to help us do that.

Dr. Johns: Dr. Tepper asks, "If the results of the cervix study were poor because of the lateral dose increase with cobalt-60, it implies that the unneeded dose caused complications leading to death. Was death by complications in cobalt-60 patients significantly increased?" As I understand it, Dr. Allt got more complications and less control of the primary with cobalt-60.

Dr. Powers: It was not primarily death due to the therapy that caused the difference in the survival rate. There were three or four deaths with the cobalt treatment, and one or more deaths with treatment by the betatron.

Dr. Johns: The best explanation for the difference is that with the high-energy radiation, more radiation goes into the tumor relative to normal tissues; that's a property of the beam.

Dr, Kramer: We have to compare the control group to a group treated in a similar way. I'm not sure that the cobalt group survival was up to the optimal cobalt-treatment, five-year results, in Stages 2B and 3.

Dr. Powers: In the Stage 3's, survival was around 25%, which was pretty worthwhile, and representative of other therapy.

Dr. Johns: Dr. Kramer says, "It has been suggested that high-energy electron beams can be shaped by magnetic fields after the beam emerges from the machine. Could you comment?" One idea being put forth by a number of people is that when the electrons come into the patient they can be curved around with magnetic fields, so that the range is very sharp. In a sense it's somewhat far-fetched because you have to have a magnet that's big enough to put the patient right between the poles. One wonders about other biological effects the magnetic field might have.

Dr. Fowler: Dr. Horn from Indianapolis says, "From your data it looks as if the regimen of five fractions for ten days of low-LET radiation is just as effective, with or without the radiosensitizing drug. Comment." What you gain from using the radiosensitizer is only the extra bit lost because of hypoxic cells; if the fractionating schedule is effective, the gain is small, in the overall time of nine or ten days.

Jean Dutreix says, "Sensitizers are likely to be less effective with a large number of fractions used in conventional fractionation, than for the three or five fractions used in your experiments with sensitizers in mice. For clinical trials would you advise keeping a standard fractionation, or reducing the number of fractions?" Conventional fractionation is probably going to be used clinically with the radiosensitizers. Since Dr. Urtason in Edmonton has started treating patients with radiation and radiation sensitizers, he has used something like three fractions a week, which is almost conventional. Radiosensitizers might make it possible and safe to use a shorter overall treatment time.

Dr. Powers: You reached a plateau of about 50% control. Was that 50% local control?

Dr. Fowler: That was 50% local control, and there were very few metastases.

Dr. Powers: How do you explain the other 50%, if increased dosage wouldn't cure them?

Dr. Fowler: It depends on the dose given, we got dose-response curves which went up nearly 100%.

Dr. Powers: So you were reaching a plateau at a normal deficit of two.

Dr. Fowler: Yes, because we chose an arbitrary level of skin damage as a parameter. If we had chosen a higher level, it would have been awfully bad for the skin reactions, but we would have cured more tumors.

Dr. Powers: Then this is a tumor in which you could have, with increase in dose, produced an increase in cure.

Dr. Fowler: Oh yes, indeed. We also found out that the use of sensitizers didn't produce any more lung metastases, one of the things we were very worried about at first.

Dr. Hall: From the limited information that we already have, with cells in culture potentially lethal damage is repaired if, after radiation, you have a suboptimal environment, such as saline. If this is so *in vivo*, then there is repair of potentially lethal damage in tumor cells (which has been demonstrated), whereas you would not expect repair in normal tissues. Therefore, potentially lethal damage adversely affects the therapeutic ratio with x-rays, whereas your model projects that it would improve that ratio.

Dr. Pohlit: You assume that there is no repair of potential lethal damage in normal healthy tissue, that the cells in such tissues have no time for repair; but they have.

Dr. Hall: The experimental evidence shows that when cells are in optimal-growth environment, they are called upon to perform intricate tasks, like division, and hence are not able to repair potentially lethal damage. Only a suboptimal environment encourages them to repair, which, *in vivo*, means within a tumor.

Dr. Pohlit: The question is whether the cells have time to repair, and they do.

Dr. Hall: Time is not the only factor, nor is it the vital factor. The suboptimal environment promotes repair of potentially lethal damage in mammalian cells; this occurs *in vivo*, as far as one can see, only in the tumor. That's why you can show potentially lethal damage repair in a mammalian tumor.

Dr. Pohlit: We have an experiment which excludes repair of normal tissue. We label the DNA of the cells with iodine-desoxyuridine; measure the gamma rays emitted by this substance from both the tumor and the healthy cells. The decrease of radioactivity measured in the tissue indicates how many cells are destroyed. In this way you can measure, as a function of time, how many cells are destroyed in the healthy tissue; within the tumor, you can measure the anoxic and oxygen-supplied cells. We don't know the precise concentration of oxygen because we have not yet developed studies for measuring this, but the studies we are doing should reveal if there is repair in healthy tissue or not. I don't know of any experiment which has excluded repair in healthy tissue, nor do I see any reason why there should not be repair. Cells in culture move too fast through the cycle to have time for repair. If you reduce environmental conditions so as to prolong cycle time, they start to repair. We see this same effect in both tumor and normal cells.

Dr. Powers: Dr. Roswit asks, "What is your philosophy in the attack on cancer—brinkmanship or least-harm?" If either an increased dosage or an additional modality is effective in a situation with greater than 10% chance of failure, if I find that the increase in complications is less than the increase in cure, I am willing to consider such increased dose or additional modality. As far as I'm concerned, in a locally curable cancer, the local persistence of the disease process is equivalent to the worst of the alternative complications.

Dr. Fowler: Why are there not more studies like Allt's on betatron treatment from somewhere in the world?

Dr. Johns: I think these studies can't be done any longer; a therapist with a betatron simply wouldn't treat a case of cervical cancer without using it.

Dr. Powers: We did a study at about the same time which compared the betatron to a method of therapy with which I disagreed strongly. Intracavitary radiation was equivalent in two series of patients with Stage 2B and Stage 3 carcinoma of the cervix. One population received additional external radiation therapy with the betatron. (The complex filtering system, as mentioned by Dr. Perez, did contribute a high increase in complications for the first three years of the study.) The other group received interstitial colloidal gold, injected into the parametrial tissue. In our institution the proponents of that method felt that it was far better than any other. In Stage 2B, both of the methods caused a 65% (crude) five years, survival with no evidence of disease. The survival in the patients receiving gold was 21%, and the survival of the patients receiving radiation externally was 51%. Our previous results for conventional radiation with cobalt and intracavitary therapy were in the range of 50% in Stage 2B. The material has not been published, but it again suggests that the betatron therapy is superior. It's a dosage phenomenon, a dose decrease in normal tissue and increase in tumor tissue. Other studies weren't done because investigators in this country were already convinced of the betatron's superiority.

Dr. Johns: Dr. Fowler, there is evidence that some sensitizers are really chemotherapeutic agents, and don't need the radiation.

Dr. Fowler: Yes, but they are not very effective chemotherapeutic agents. In our experiments in solid tumors, one or two days' delay in regrowth comes from use of the drugs by themselves, but the delay extends to as much as 40 days with the radiation added while the drug is present. There's a lot of work to be done on the mixing of sensitizers with various chemotherapeutic agents, but these sensitizing drugs do not deplete the bone marrow, a main side effect of most cytotoxic agents.

Dr. Fowler: Arthur Boyer from Massachusetts General Hospital, Boston, asks, "Could you get large enhancement ratios with radiosensitizers if you injected 0582 intralesionally, just before irradiation?" There is work to be done on the route of administration. We administer by mouth in patients, and intraperitoneally in mice; the routes are rather similar. Richard Johnson in Buffalo is administering intravenously; we think that's probably quicker. But it takes some time, a number of minutes, for these substance to diffuse to all the cells. Even if you put the stuff directly into the tumor, you've still got to find out how long it takes to reach the hypoxic cells and become effective.

Dr. Leeper from Jefferson asks: "How far will Flagyl or 0582 actually diffuse through metabolizing tissue?" All the way, as far as we know. The evidence from the single dose curve indicates that all the hypoxic cells have been sensitized. There is no real physiochemical reason why this should not happen, if the substance will get into cells and diffuse through tissue. It takes a little time. Dr. Leeper also asks, "During a fractionated regimen of radiation therapy, would the electron-affinic agents be as effective in the long run if administered during the latter part of radiation treatment, after the tumor shrinks?" Yes; if these drugs turn out to be very toxic, then one would just use them in the latter half of the course of radiotherapy. Another question from Dr. Leeper, "0582 did not enhance local control by very much, with certain of the fractionation intervals, when the optimum 48-hour interval was used. Should we continue to study the rates and parameters of reoxygenation in tumors, or will the use of electron-affinic agents remove the need for specific time-dose relationships which take advantage of tumor-specific rates of reoxygenation?" It's difficult to find out what factors effect optimum reoxygenation, even in mouse tumors where you have a lot of time and information. The radiosensitizers might help us to bypass this complex problem.

Dr. Fowler: Dr. Fish of Stanford asks about hyperthermia. Hyperthermia seems to have some extemely interesting effects. Within the last year or two there have been some very good experiments done, and there's a strong basis for more investigation. It is exciting clinically, and has gone into use without laboratory backup in one or two places. Studying hyperthermia must be rather like looking at the effects of radiation in the 1920's; you can get contradictory results from different laboratories, because there are so many unknown parameters.

Dr. Powers: How does one produce a given gradient of hyperthermia? Dr. Pohlit, what influence would hyperthermia have on your model of the processes, and what do you think of hyperthermia as an enhancer of radiation therapy effects?

Dr. Pohlit: Like radiosensitizers, in a high dose hyperthermia is very dangerous. It is not easy to use really high temperatures in a patient, but it should be done.

Dr. Powers: Well a number of years ago in Germany this method was being used by itself, without radiation therapy, for control of tumors. I wonder what has happened to that experience, and other similar ones?

Dr. Pohlit: These are the experiments now being checked in East Germany. In West Germany some hospitals started this but closed the experiments due to failure in certain patients. There is now no hospital in West Germany which is trying this method.

Index

(Page numbers in italics refer to figures or tables.)

Attenuation, coefficients, 8
   electrons, 158
   50% layer, 210

Beam characteristics, ideal, 351
Betatron, 5, 19, 53
Biologic effect, 43 44, 45
Bladder, 21, 74, 111, *111*, *112*, 122, 125, 295
Box technique, 22, 30, *34*, 108, 230, *231*, 335, 336
Brain, 122, 125
Breast, *54*, 65, *66*, *67*, 73, 81, 118, 199, 200
Buccal mucosa, 180, 181
Build-up depth, 20, 107, 207

Carcinogenesis, 353
Cell cycle, 323
Cell survival curves, 41, *42*, *46*, 77–79, 80, 117, 199, *313*, *324*, *326*, *328*, *329*
Cervix, 5, 104, 107, 108, 111, 118, 198, 230–235, *336*, 359
Chest wall, 191, 199, 200
Complications, *228*, *229*, *235*, *238*, *238*, 350, 352
   lung fibrosis, 272
   osteonecrosis, 303, 304
   cartilage necrosis, 304
   pulmonary function, 306, 307
   radiation pneumonitis, 272
   xerostomia, 22, 304
Computerized transaxial tomography, 92, 200
Contours, 89, 92
Continuous irradiation, 125

Depth dose, 6, 19, *19–21*, 53, *54*, *62*, *63*, 104–107, *172*, *181*, 218, 220, 335 *339*
   electrons, 53
Dose distribution, 21, *23*, *26*, *28*, *29*, 56, 57, *64*, *70*, *75*, *94*, *95*, *109–112*, *173*, *224*, *225*, *230–232*, *236*, *257–259*, *336–338*, *342*, *343*
   volume, 22
   precision, 87
Dose fractionation, *324*
Dose measurements, in-vivo, 95, 160
   gonadal, 96–97
   electrons, 139
   film, 142
   ionization chambers, 205
Dose optimization, 238
Dose rate, 42–45, *47–49*, 78, 79, 82, 117, 335
Dose response curve, 350

Electron affinity, 312
Electrons, 22, 53, 131
   electivity, 60
   range, 133
   patching fields, 166
   clinical applications, 171
   advantages, 190
Endometrium, 31
Equivalent dose, *120*
Esophagus, 26, 28, 113, 122, 197
Ethmoid, 24, *28*, *29*,

Fibrosis, 30, 31, 59, 175, 353
Flattening filter, 7, 108, 219
Fractionation, 39–41, *43*, *44*, 78, 89, 117, *119*, *121*, *123*, 197, 199, 220, 260, 261, 314, 316, 358, 360

Genetic perturbation, 353
Gonadal dose, *96, 97*
Growth retardation, 353

Head and neck, 22, 65–67, 224
Hemoptysis, 263
Heterogeneity, 53, 93, 173, 212
  corrections, 245
Hyperthermia, 360
Hypoxia, 39–40, 79, 80, 82, 181, 220, *311*, 312, *313*, 314, 327, 354, 359

Immune response, 220
Immunosuppression, 353
Inhomogeneity, 53, 78, 200, 301, 306
Integral dose, 221, 223, 292, 304–306, 339–340
Intracranial lesion, 55
Intracavitary shield, 54
In-vivo dosimetry, 95
Irreparable damage, 325, 331
Isodose curves, 53

Jejunal crypt cells, 43–44

Kidney, 295

Larynx, 118, 119, 304
LET, 103
Linacs, 7
Lip, *189, 190*
Lung, 118, 123, *257–259, 264–271, 273, 274,* 275
Lung cancer with tuberculosis, 262, 263
Lung abscess, 263

Maxillary sinus, *59, 60*
Mediastinum, 26, 73
Melanoma, 117
Middle ear, 65, *68*
Mixed beam, 24, *31*, 32, 75, 78, 84, 171, 174, 175, 295, 296, 350
Multifraction, 124, 125
Mucositis, 31

Nasal cavity, 24, *59, 60, 183–187*

Nasopharynx, 22, *59, 60, 192,* 224, 226, 227, *235*
Neck, 188
Neutrons, 40, 103
  contamination, 77
Nitroimidazoles, 314, 316, 318, 321
Nodes, 22, 67, 80, 180, 181, 191
Nominal Standard Dose (NSD), 118, 119
Nose, 67, 181

Oxygen enhancement ratio (OER), 40
Oral cavity, 22, 73, 181
Oropharynx, 22, 73

Pancreas, 83, 84
Parotid, 67, 190, *193, 194*
Pelvis, 30
Penumbra, 53, 106, *107,* 211
Pharyngeal wall, 24
Pharynx, 118
Photons, 105
  physical characteristics, 217
  advantages, 221
  disadvantages, 221
Potentially lethal damage, 39, 323, 325, *326, 327,* 329, 358
Prostate, 30, *222, 223,* 233, 234, 236, 237

Radiosensitizers, *316, 317*
RBE, 42, 58, 77
Reassortment, 39
Reoxygenation, 39, 117, 352
Repair, 39, 45
Repopulation, 39, 117
Rets, 59, 60
  NSD, 119

Scanning, electron beams, telecentric, 289
  clinical examples, 295
Scattering foils, 14, 149, 287
Sensitizers, 40, 311, 359, 360
Shielding, 54
Single dose, 122, 260–261
Sinuses, 67

Skin, *178–180*, 181, 197
  fibrosis, 32, 59
  radiation reaction, 118, 171, 172, 174, 201, 202, 314
  sparing, 104, 105
  tolerance, 58
Split course, 124, 352
Subcutaneous fibrosis, 32, 59, 171, 202
Sublethal damage, 39, 42–45, 199, 323, 328–330
Surface dose, 12, 217, *218*, *219*
Survival curve, 41, 42, 60, 323, 329
Survival data, cervix, *235*
  lung, *262*, 272, 275–280, 308
SVC, 266

Targets, 7
Therapeutic ratios, 42
Thorax, 28
Thyroid, 66, *69*, *70*, *228*
Time-dose relationships, 117, 260, *261*
Tissue homogeneity, 221
Tonsil, *23–25*, 123, 184, 224, 225–229
Treatment planning, 87
  computerized, 98
Tumor cell factors, 219, 220
Tumor dose, 87
Tumor localization, 349
Tumor volume, 87, 89, 220

Ultrasonography, 90, 91

X-ray contamination, 13, 15

Yield, x-ray, Z-dependence, 9